世界技能大赛
印刷媒体技术项目专业英语

李不言 王一青 管雯珺 编著

SHIJIE JINENG DASAI YINSHUA MEITI
JISHU XIANGMU ZHUANYE YINGYU

文化发展出版社
Cultural Development Press
·北京·

图书在版编目（CIP）数据

世界技能大赛印刷媒体技术项目专业英语 / 李不言，王一青，管雯珺编著. -- 北京：文化发展出版社，2024.11. -- ISBN 978-7-5142-4453-3
Ⅰ.TS801.8
中国国家版本馆 CIP 数据核字第 2024TG3100 号

世界技能大赛印刷媒体技术项目专业英语

李不言 王一青 管雯珺 编著

出 版 人：宋　娜	
责任编辑：李　毅　雷大艳	责任校对：侯　娜
责任印制：邓辉明	封面设计：魏　来

出版发行：文化发展出版社（北京市翠微路 2 号 邮编：100036）
发行电话：010-88275993　010-88275710
网　　址：www.wenhuafazhan.com
经　　销：全国新华书店
印　　刷：中煤（北京）印务有限公司
开　　本：787mm×1092mm 1/16
字　　数：304 千字
印　　张：12
版　　次：2024 年 11 月第 1 版
印　　次：2024 年 11 月第 1 次印刷
定　　价：58.00 元
ＩＳＢＮ：978-7-5142-4453-3

◆ 如有印装质量问题，请与我社印制部联系。电话：010-88275720

前　言

世界技能大赛（WorldSkills Competition，WSC）是由世界技能组织（WorldSkills International，WSI）主办的最高层级的世界性职业技能大赛，被誉为"世界技能奥林匹克"。

世界技能大赛中的印刷媒体技术（Print Media Technology）项目于2005年第38届世界技能大赛时首次设立。目前，我国选手已连续参加了第42届、43届、44届、45届和2022年世界技能大赛特别赛共5届印刷媒体技术项目比赛，并取得一金一银一铜和一个优胜奖的优异成绩，我国成为该项目首个实现各类奖牌"大满贯"的国家。

世界技能大赛印刷媒体技术项目包括胶印（Offset Printing）、数字印刷（Digital Printing）和附加任务（Additional Task）三大模块，模块下设20余项任务。参赛期间，选手需具备独立阅读施工单（Job Ticket），以及与专家、选手进行技术交流和日常交流的英语能力。

《世界技能大赛印刷媒体技术项目专业英语》按照世界技能大赛印刷媒体技术项目选手参赛所需，共包括施工单、对话、竞赛口语和练习四个部分。第一部分是施工单，精选了第44届、45届比赛中的15个任务真题，覆盖了印刷媒体技术项目三大模块的技能需求，选手通过真题练习，可以全面掌握高频专业词汇，熟悉施工单的表达方法，具备独立阅读施工单的能力。第二部分是对话，作者基于多年参加世界技能大赛的经验，编写了10篇日常交流对话，内容包含人员介绍、世界技能大赛文化等，为选手参赛期间与其他国家人员沟通交流打下良好的基础。第三部分是竞赛口语，主要根据世界技能大赛设备熟悉日和竞赛日两个重要时间节点所需，结合印刷媒体技术项目的特点，模拟还原了选手有可能向专家提出的问题，为选手顺利参赛做好语言铺垫。通过第二、第三部分的学习，选手可以基本掌握参赛期间与专家、选手进行技术及日常交流的英语能力。第四部分是练习，主要用于教学过程中对第一部分施工单中的词汇及句子掌握水平的评估。《世界技能大赛印刷媒体技术项目专业英语》适用于各院校参赛选手英语能力的全面培养，同时可作为印刷相关专业的专业英语课程教材或参考书籍来使用。

《世界技能大赛印刷媒体技术项目专业英语》由上海出版印刷高等专科学校教师李不言、王一青、管雯珺编写。李不言曾担任第 44 届、45 届世界技能大赛和 2022 年世界技能大赛特别赛印刷媒体技术项目的中国技术指导专家组组长，中华人民共和国第一届、第二届职业技能大赛印刷媒体技术项目的裁判长，主要负责本书第一、第四部分的编写。王一青曾担任第 45 届世界技能大赛和 2022 年世界技能大赛特别赛的中国翻译，主要负责本书第二、第三部分的编写。管雯珺曾担任中华人民共和国第一届、第二届职业技能大赛印刷媒体技术项目的裁判长助理，主要负责本书附录一扩展阅读、附录二词汇表的编写及全书统稿。

由于作者水平有限，恳请印刷行业各位前辈、专家对本书的不足之处予以批评指正。

编者

2024 年 5 月

目 录

Part 1 Job Ticket 1

Chapter 1 Offset Printing 2
Unit 1 Post Cards 2
Unit 2 Feeder / Delivery Set Up 10
Unit 3 Image and Color Registration 14
Unit 4 Four Color Press Operation 18

Chapter 2 Digital Printing 23
Unit 5 Tent Cards 23
Unit 6 *Vision 2025* Booklet 29
Unit 7 Preflight and Edit Digital Image 34
Unit 8 Preflight 37
Unit 9 Engineer Workflow and Print a Digital File 42
Unit 10 Color Correct and Print a Digital File 45

Chapter 3 Additional Task 48
Unit 11 Maintenance 48
Unit 12 Defect Detection 51
Unit 13 Fault finding, maintenance 57
Unit 14 Ink Mixing 63
Unit 15 Ink Verification 67
Unit 16 Cutting Exercises 71
Unit 17 Cutting Post Cards 74
Unit 18 Simulation 77

Part 2 Dialogue 81

    Unit 1 Who Is Who? 82

    Unit 2 WSC Culture 87

    Unit 3 Skills Management Plan 93

    Unit 4 Meeting New Friends 97

    Unit 5 If You've Got Any Question? 101

    Unit 6 Dispute Resolution 104

    Unit 7 How to Get to the Competition Venue? 108

    Unit 8 Publication of Results 113

    Unit 9 Medals and Awards 117

    Unit 10 Doing an Interview 120

Part 3 Oral English in Competing Days 125

    Unit 1 Familiarization Day 126

    Unit 2 Competition Day 128

Part 4 Exercise 131

    Unit 1 Offset Printing 132

    Unit 2 Digital Printing 137

    Unit 3 Additional Task 142

Appendix 1 Expanded Reading 149

Appendix 2 Vocabulary & Phrase 163

Part 1

Job Ticket

Chapter 1 Offset Printing

Unit 1 Post Cards

Description

This job is Printed Offset and will run 4/4 front and back using the work and turn imposition — competitors to punch, bend, mount plates, establish registration and pre-set the press for high quality and productivity, saving the required number of press sheets. Turn the job and print the second side so that the front to back registration is correct and color meets density requirements with no marking. Wash up the units (rollers, plates, blankets and impression cylinders with automatic program) and clean wash-up blades.

Job	Post Cards, 4/4
Number of copies	600
Set-up waste	500
Total amount of paper	1100
Colors	4/4 – CMYK Process /CMYK Process
Paper size	Sheet size 350 mm×500 mm
Finished size	Not Applicable
Paper	250 g/m^2 Gloss Coated
Ink	C, M, Y, K
Machinery (Printing Press)	Heidelberg Speedmaster SM 52-4

Part 1 Job Ticket

Job status Content proof and target density values supplied at press

Available time for the total task 2 hours

During the printing

FIRST PRINTED SIDE

Competitors will achieve color and registration, provide one sheet to the experts for quality verification, and stop production. Upon receiving approval from experts that the sheet is 'OK sheet', competitors will set counter to zero, turn on the counter, turn in make ready sheets, and continue press run with empty trolley.

Competitors will pull one sample sheet for evaluation during printing at 300, and 600 (± 50) sheets printed. Identify one sample from the numerous sheets. Immediately give the sample pull to the experts for color scanning.

Competitors will be evaluated against posted density value.

Competitors will turn all the first side finished printed sheets and make ready for second side.

SECOND PRINTED SIDE

Competitors will achieve color and registration front and back, provide one sheet to the experts for quality verification, and stop production. Upon receiving approval from experts that the sheet is 'OK sheet', competitors will set counter to zero, turn on the counter, turn in make ready sheets, and continue press run with empty trolley.

Competitors will monitor color and register through press run. Make a pull at 500 sheets (± 50) and give to the experts for color scanning.

After printing	Competitors will leave delivery trolley in the press. The competitors must clean the press with the automatic program (printing units, blanket and impression cylinder) and manual cleaning of the 4 wash-up blades in correct and safe ways.
Additional knowledge	The X-rite technicians will print a report of color density and give it to the experts.

The job ticket above comes from task 1.1 of the Print Media Technology project of the 44th WSC.

▶ [VOCABULARY & PHRASE]

offset / ˈɔːfset / *n.* 胶印 *adj.* 胶印的，平版印刷的 offset printing 胶印	bend / bend / *v.* 弯（版） bend in 使……往里弯 bend over 俯身；折转
imposition / ˌɪmpəˈzɪʃ(ə)n / *n.* 拼版 turn imposition 翻面	mount / maʊnt / *v.* 安装（印版）；组织；增多，上升；增强 mount up 增长；上升
punch / pʌntʃ / *v.* （印版）打孔 *n.* 一拳，一击 punch press 打孔机 punch card 打孔卡	plate / pleɪt / *n.* 印版；盘子，碟子；一盘（食物） fruit plate 水果盘 registration / ˌredʒɪˈstreɪʃ(ə)n / *n.* 套准；定位；登记，注册；挂号

registration form 注册表

press / pres /
n. 印刷机；新闻工作者；新闻报道
v. 压，挤，推
printing press 印刷机

quality / ˈkwɑːlətɪ /
n. 质量
adj. 优质的，高质量的
quality control 质量控制
high quality 高质量

productivity / ˌprɒdʌkˈtɪvətɪ /
n. 生产率，生产力
labor productivity 劳动生产率

production / prəˈdʌkʃn /
n. 生产，制造
production line 流水线

sheet / ʃiːt /
n. 纸张，纸片
OK sheet OK 样
sheet size 纸张规格
make ready sheet 校版纸

unit / ˈjuːnɪt /
n.（印刷机）单元；单位

unit of time 单位时间

roller / ˈroʊlər /
n. 辊；滚动（或碾压）东西的人
roller coaster 过山车

blanket / ˈblæŋkɪt /
n. 橡皮布；毯子；覆盖层；气氛
v. 覆盖；消除（声音）
adj. 总括的，全面的
blanket order 总订单

impression / ɪmˈpreʃ(ə)n /
n. 压印，印记；印次
impression cylinder 压印滚筒

cylinder / ˈsɪlɪndər /
n.（印刷机）滚筒；圆柱体，圆筒；汽缸
gas cylinder 煤气罐

automatic / ˌɔːtəˈmætɪk /
adj. 自动的
automatic control 自动控制

program / ˈproʊɡræm /
n. 程序
v. 设置
automatic program 自动程序

blade / bleɪd /

n. 刀片，刀刃

wash-up blade 刮墨刀

applicable / ˈæplɪkəbl /

adj. 适用的，适当的

not applicable 不适用

gloss / glɑːs /

n. 光彩；光泽涂料；注释

gloss coated 高光涂布

gloss oil 上光油

status / ˈsteɪtəs; ˈstætəs /

n. 状况，情形

job status 工作状态

content / ˈkɑːntent /

n. 内容；目录；所含物

adj. 满足的

v. 使满意，使满足

free content 免费内容

water content 含水量

content provider 内容提供者

proof / pruːf /

n. 样张，校样

content proof 样张

target / ˈtɑːrɡɪt /

n. 目标，指标；（攻击的）对象；靶子

v. 把……作为目标；面向，把……对准（某群体）

target market 目标市场

on target 切中要害

density / ˈdensəti /

n. 密度；稠密，密集

energy density 能量密度

value / ˈvæljuː /

n. 价值；等值

target density values 目标密度值

verification / ˌverɪfɪˈkeɪʃn /

n. 验证；核查；证实

quality verification 质量检验

ink verification 油墨验证

counter / ˈkaʊntər /

n. 计数器

adj. 反面的，对立的

loop counter 循环计数器

trolley / ˈtrɑːli /

n. 收纸台板

v. 用手推车运

trolley case 拉杆箱

sample / ˈsæmpl /
n. 样张；样本，样品
v. 品尝；体验（活动）；对……作抽样调查
sample size 样本尺寸

scanning / ˈskænɪŋ /
n. 扫描
color scanning 色彩扫描

monitor / ˈmɑːnɪtər /
n. 监控器，显示器；班长
v. 监视；监听
system monitor 系统监视器

delivery / dɪˈlɪvərɪ /
n. 收纸；递送，投递；分娩
time of delivery 交货时间

technician / tekˈnɪʃn /
n. 技术人员，技师
X-rite technician 爱色丽技术人员

front and back 正面和反面
number of copies 印张数量
set-up waste 校版纸
pre-set 预设
total amount of paper 纸张总数
paper size 纸张规格
finished size 成品规格
Heidelberg Speedmaster 海德堡速霸
turn on 打开
turn in 上交
available time for the total task 工作总时间
first printed side 第一印刷面

▶ [TRANSLATION]

第1单元　明信片

描述

这个工作是4/4正反面自翻胶印工作。选手通过印版打孔、弯版、安装印版、套准及预设印刷机来获得高质量和高生产率，节省所需数量的印刷纸张。将印刷品翻面以印刷第二面，确保正反面套准正确，色彩达到密度要求且无污点。清洁印刷机单元（使用自动程序清洁辊、印版、橡皮布和压印滚筒）和刮墨刀。

工作	明信片，4/4
印张数量	600
校版纸	500
纸张总数	1100
色彩	4/4 – CMYK /CMYK
纸张规格	350 mm×500 mm
成品规格	不适用
纸张	250 g/m^2 高光涂布纸
油墨	C，M，Y，K
机器（印刷机）	海德堡速霸 SM 52-4
工作状态	通过印刷机提供样张和目标密度值
工作总时间	2 小时
印刷中	**第一印刷面** 选手（认为）色彩和套准达到要求后，提交一张印张给专家进行质量检验，并停止生产。当专家认可提交的印张为"OK 样"后，选手将计数器归零，打开计数器，上交校版纸，使用空的收纸台板继续印刷。 选手在印刷至第 300，600（±50）张处时提交一张样张用于评估。从众多印张中确定一张。并立即将样张交给专家进行色彩扫描。 选手提交的样张的密度值将被评估。选手将所有完成第一面印刷的印张翻面并准备第二面印刷。

第二印刷面

印刷中	选手（的印刷样张）达到色彩和正反面套准要求后，提交一张印张让专家进行质量检验，并停止生产。直到专家认可提交的印张为"OK 样"后，选手将计数器归零，打开计数器，上交校版纸，使用空的收纸台板继续印刷。 选手在印刷过程中进行色彩和套准监控。在印刷至第 500（±50）张处时提交一张样张让专家进行色彩扫描。
印刷后	选手将收纸台板留在印刷机里。 选手必须使用自动程序清洁印刷机（印刷单元、橡皮布和压印滚筒），并以正确、安全的方式手动清洁 4 个刮墨刀。
附加知识	爱色丽的技术人员将打印一份色彩密度报告，并将其提交给专家。

以上施工单选自第 44 届世界技能大赛印刷媒体技术项目的任务 1.1。

Unit 2　Feeder / Delivery Set Up

Description

The primary task of all offset press operators is to make sure the feeder and delivery are set properly, so there will be no double sheets, damaged sheets, feeder stops and delivery problems.

Marking

Judgment and measurement assessment will be used.

Time allotment

50 minutes

Task

1. When the judge says "begin", load <u>1000 sheets of 80 gsm gloss 350 mm×500 mm</u> paper into feeder of Heidelberg press.

2. Adjust feeder and delivery controls so paper will flow through press.

(a) Follow the Heidelberg procedures as defined at Wiesloch training session.

(b) Sheets should have <u>NO</u> damage, marking, or defect in the finished paper stack.

(c) There should be <u>NO</u> feeder or delivery fault stop during task run.

(d) Press speed must run at a <u>minimum 12000 SPH</u>.

(e) Competitors may choose to run at full speed of 15000 SPH for additional marks.

3. Deliver <u>700 sheets</u> perfectly to the delivery of the press.

4. <u>Do not remove</u> the finished stack from the press or the remaining sheets from the feeder.

5. Notify the judge when you have completed the task.

The job ticket above comes from task 1.1 of the Print Media Technology project of the 45th WSC.

[VOCABULARY & PHRASE]

feeder / ˈfiːdər /
n. 飞达；支线；喂食器；奶瓶；饲养员
feeder stop 飞达停止

operator / ˈɑːpəreɪtər /
n. 操作员；经营者
system operator 系统操作员

judgement / ˈdʒʌdʒmənt /
n. 评价（评分）；意见；判断力；审判
subjective judgement 主观判断

measurement / ˈmeʒərmənt /
n. 测量（评分）；三围；计量；测量单位
unit of measurement 计量单位
measurement data 测量数据
measurement error 测量误差

assessment / əˈsesmənt /
n. 评判，评价
assessment method 考核方法
market assessment 市场评估

task / tæsk /
n. 任务
v. 派给……任务
task analysis 任务分析

load / loʊd /
v. 装入（纸张）；上（子弹）；载入（计算机程序）
n. 负载，重荷
full load 满载

marking / ˈmɑːrkɪŋ /
n. 脏污，留下痕迹；标识
v. 做记号
marking pen 记号笔
marking tool 画线工具

defect / ˈdiːfekt /
n. 弊病；缺陷，毛病
v. 背叛，叛变
zero defect 零缺陷
surface defect 表面损坏

remove / rɪˈmuːv /
v. 搬出；移开；废除；把……免职；脱下
remove grain 移除颗粒
remove all 全部删除

notify / ˈnoʊtɪfaɪ /
v. 示意；通报，告知；申报
notify party 通知方

allotment / əˈlɑːtmənt /
n. 分配；配额
time allotment 时间分配

stack / stæk /
n. 堆；大量，许多
a stack of 一堆……

gsm 克 / 平方米，gram per square metre 的缩写，与 g/m² 意思相同
SPH 张 / 小时，sheets per hour 的缩写

set up 设置
double sheets 双张
damaged sheet 纸张损坏
Heidelberg procedure 海德堡（工作）流程
finished paper stack 收纸堆
fault stop 故障（造成的）停止
press speed 印刷速度
full speed （印刷机）最大印刷速度
additional mark 附加分
remaining sheets 剩余纸张

▶ [TRANSLATION]

第 2 单元　飞达 / 收纸设置

描述
所有胶印机操作员的首要工作是确保飞达和收纸正常，这样才能避免双张、纸张损坏，飞达停止和收纸故障。

评分
将使用评价评分和测量评分进行评判。

时间分配
50 分钟

任务
1. 当裁判宣布开始后，将 1000 张 350 mm×500 mm 的 80 g/m² 高光纸装入海德堡印刷机的飞达。

2. 调节飞达和收纸控制部件，使纸张在印刷机中输送。

(a) 遵守在维斯洛赫（德国）培训时的海德堡工作流程。

(b) 输送的纸张在收纸堆中须无损坏、无痕迹、无收纸缺陷。

(c) 输纸过程中不能出现因故障造成的飞达或收纸停止。

(d) 最小输纸速度为 12000 张 / 小时。

(e) 若选手选择使用最大（印刷）速度 15000 张 / 小时将获得附加分。

3. 将 700 张纸完美输送到印刷机收纸处。

4. 不要将收纸堆从印刷机上搬出，也不要将飞达上剩余的纸张取出。

5. 当你完成任务时示意裁判。

以上施工单选自第 45 届世界技能大赛印刷媒体技术项目的任务 1.1。

Unit 3 Image and Color Registration

Description

One of the most important task of all offset press operators must perform is to assure the image is in the correct position on the sheet and that all colors are in perfect registration.

Marking

Judgment and measurement assessment will be used.

Time allotment

50 minutes

Task

1. When the judge says "begin", you will need to assess the position of the images on the press sheet by printing an inspection sheet.

You will be provided with <u>1000 sheets of 115 gsm satin</u> grain long paper.

2. Make the needed adjustments to any part of the press to bring the image into the correct position.

(a) The image should be correctly positioned on the paper.

(b) All colors should be in perfect registration.

(c) Use the ink key settings as you find the press. The Heidelberg technician will have the color zones within standard — **YOU DO NOT NEED TO ADJUST COLOR**.

Cyan	1.30 ± 0.09
Magenta	1.30 ± 0.09
Yellow	1.15 ± 0.09
Black	1.60 ± 0.12

3. When you have the position and registration correct — give your OK sheet to the judges.

(a) Your OK sheet will be evaluated for image position on the sheet.

(b) Your OK sheet will be evaluated for registration.

(c) If your OK sheet's color zones closest to edge of sheet are not within standard, it will not be accepted.

4. Deliver 700 sheets perfectly to the delivery of the press.

5. If you feel you need to make additional adjustments to registration, you may do so.

6. Do not remove the finished sheets from the press or any of the remaining sheets from the feeder.

One press sheet — the 30th sheet from the top of the stack will be evaluated for final registeration.

7. Notify the judge when you have completed the task.

There is No evaluation for color on this task — **ONLY POSITION AND REGISTRATION**.

The job ticket above comes from task 1.2 of the Print Media Technology project of the 45th WSC.

▶ [VOCABULARY & PHRASE]

image / ˈɪmɪdʒ /
n. 图像；形象，印象
v. 作……的像，描绘……的形象
image sensor 图像传感器
digital image 数字图像
virtual image 虚拟图像

position / pəˈzɪʃn /
n. 位置；地点
v. 安置；为（产品、服务、业务）打开销路
social position 职位；社会地位

assess / əˈses /
v. 评估；征税，处以罚金
assess risks 资产风险；风险评估

inspection / ɪnˈspekʃn /
n. 检查；视察
inspection sheet 校准印张
inspection certificate 检验证书

satin / ˈsætn /
n. 缎子；缎子衣服
adj. 光滑的
satin grain long paper 光面长丝缕纸张

grain / greɪn /
n. 纹理；谷物；颗粒；（表面的）质地
v. 使表面（或纹理）粗糙；成粒状
add grain 添加杂点

adjustment / əˈdʒʌstmənt /
n. 调整；调节；调节器
seasonal adjustment 季节性调整
structural adjustment 结构调整
price adjustment 价格调整
speed adjustment 速度调节

standard / ˈstændərd /
n. 标准，水平，规范
adj. 普通的，标准的

standard error 标准误差

cyan / ˈsaɪən; ˈsaɪæn /
n. 青色
adj. 青色的

magenta / məˈdʒentə /
n. 品红；洋红
adj. 品红色的；洋红色的

ink key （胶印机）墨键
color zones 色彩（检测）区
edge of sheet 纸张边缘
finished sheets （印刷）完成的印张
top of the stack 收纸堆顶部

▶ [TRANSLATION]

第 3 单元　图像和色彩套准

描述

所有胶印机操作员必须执行的一项重要任务是保证图像在印张上的位置正确，以及所有色彩套准完美。

评分

将使用评价评分和测量评分进行评判。

时间分配

50 分钟

任务

1. 当裁判宣布开始后，你需要印刷一张校准印张来评估图像在印张上的位置。

你将会得到 1000 张 115 g/m² 的光面长丝缕纸张。

2. 对印刷机的任何部分进行必要的调节，使图像处于正确的位置。

(a) 印张上的图像位置需正确。

(b) 所有色彩需完美套准。

(c) 使用印刷机上已经设置好的墨键。海德堡的技术人员已经将色彩（检测）区色彩调节至标准范围内——你无须对颜色进行任何调整。

青	1.30 ± 0.09
品红	1.30 ± 0.09
黄	1.15 ± 0.09
黑	1.60 ± 0.12

3. 当图像位置和套准都调节正确后，向裁判提交你的 OK 样张。

(a) 你提交的 OK 样张的图像位置将被评估。

(b) 你提交的 OK 样张的套准将被评估。

(c) 你提交的 OK 样张中位于纸张边缘的色彩（检测）区若不符合标准，将不会被接受。

4. 完美输送 700 张纸张至印刷机收纸处。

5. 如果你认为套准（在印刷过程中）需要进一步调节，可以进行调节。

6. 不要从印刷机上搬出印刷完成的印张，也不要将飞达上剩余的纸张取出。

从收纸堆顶部向下数第 30 张的这张印张将被用于最终的套准评判。

7. 当你完成任务时示意裁判。

本任务不对色彩进行评估，**只评估（图像）位置和套准**。

以上施工单选自第 45 届世界技能大赛印刷媒体技术项目的任务 1.2。

Unit 4 Four Color Press Operation

Description

The press operators must be able to do their jobs even with some information missing.

Marking

Judgment and measurement assessment will be used.

Time allotment

2 hours

Task

This job is Printed Offset. Competitors bend, mount plates, select the correct paper, establish registration and pre-set the press for high quality and productivity, save the required number of press sheets.

1. When the judge says "begin", select the correct paper for the job from the table and identify correct plates for the printing unit.

   (a) Choose 1000 sheets of paper A or B.

   (b) Select the correct color printing plate by placing in in front of either the C, M, Y, or K printing unit.

2. Set the feeder and delivery for faultless feeding and perfect delivery.

3. Bend and install the plates on the <u>correct printing unit</u>.

4. Position the image correctly on the paper and <u>register all 4 colors perfectly</u>.

5. Bring all color spots in the image to the correct density.

   (a) Cyan 1.30 ± 0.09

   (b) Magenta 1.30 ± 0.09

   (c) Yellow 1.15 ± 0.09

   (d) Black 1.60 ± 0.12

6. <u>Measure color</u> of your press sheets.

(a) Give your "OK" sheet to the judge.

(b) You will be marked for your registration and color.

7. Print 500 sheets and deliver a perfect stack.

(a) Measure color of your last sheet printed and give to the judges.

(b) You will be marked for registration and color of the 30th sheet from the top of the stack.

8. Do not remove the stack from the press or any unused sheets from the feeder.

9. Clean the press for the next job — wash rollers, blankets, wipe the roller wash-up blades. Clean and organize the work area.

Do not remove the ink from the fountain.

10. Notify the judges when you have completed the task.

The job ticket above comes from task 1.3 of the Print Media Technology project of the 45th WSC.

▶ [VOCABULARY & PHRASE]

information / ˌɪnfərˈmeɪʃ(ə)n /	faultless responsibility 无过失责任制
n. 信息；消息；资料；情报；信息台	
information system 信息系统	feeding / ˈfiːdɪŋ /
information technology 信息技术	n. 输纸；饲养
basic information 基本信息	feeding system 控食系统
establish / ɪˈstæblɪʃ /	install / ɪnˈstɔːl /
v. 建立；证实	v. 安装；正式任命；安顿
establish its own 自立门户	easy to install 安装灵活
faultless / ˈfɔːltləs /	spot / spɒt /
adj. 完美的；无缺点的	n. 斑点；地点；（人体的）部位；污渍；（皮肤上的）丘疹；排名位置；几滴（液体）；
faultless operation 无事故运行	

困境；现金交易；（人格或名誉的）污点；聚光灯（spotlight 的简称）
v. 看见；注意到；（对比赛对手）让分；使有污迹；将（台球）放在置球点上
adj. 现货交易的，立即支付的
color spots 色块
spot color 专色
on the spot 当场
color bar spot 色带（上的）色块

measure / ˈmeʒər /
v. 测量
n. 措施，办法
technical measure 技术措施

mark / mɑːrk /
n. 污点；标志；记号；分数
v.（给学生或功课）打分；做标记；标示；纪念；标志（重要事件或时刻）；（特征，特点）标志；留心；盯住（对手）；分隔
trade mark 商标

unused / ˌʌnˈjuːzd; ˌʌnˈjuːst /
adj. 不用的；从未用过的
unused land 荒地

wipe / waɪp /
v.（用布、手等）擦干净；解雇（某人）；刷（卡）
n.（湿）抹布；擦
wipe off 擦掉；还清
wipe out 擦净

organize / ˈɔːrɡənaɪz /
v. 组织；安排；照料
organize my thoughts 整理思绪

fountain / ˈfaʊnt(ə)n /
n.（胶印机）墨斗；喷泉；天然泉
musical fountain 音乐喷泉

four color press 四色印刷机
place in 放置
work area 工作区域

▶ [TRANSLATION]

第4单元　四色印刷机操作

描述

在缺失一些信息的情况下，印刷机操作员也要能够进行工作。

评分

将使用评价评分和测量评分进行评判。

时间分配

2 小时

任务

这个工作是胶印。选手弯版、安装印版、选择正确的纸张、套准和预设印刷机，以获得高质量和高生产率，节省所需数量的印刷纸张。

1. 当裁判宣布开始后，根据本工作从桌上选择正确的纸张，为印刷单元选择正确的印版。

(a) 从 A、B 两种纸张中选择一种（A、B 各 1000 张）。

(b) 选择正确颜色的印版，将印版（按颜色）分别放置在 C、M、Y、K 印刷机组前。

2. 对飞达和收纸进行调节，保证输纸、收纸完美。

3. 弯版，将印版安装至<u>正确的印刷单元内</u>。

4. 保证纸张上图像位置正确，<u>4 色套准完美</u>。

5. 将图像上所有色块调节至正确的密度。

(a) 青 1.30 ± 0.09

(b) 品红 1.30 ± 0.09

(c) 黄 1.15 ± 0.09

(d) 黑 1.60 ± 0.12

6. 对你的印张进行<u>色彩测量</u>。

(a) 将"OK 样"提交给裁判。

(b) 你的样张套准和色彩将被评分。

7. <u>印刷 500 张印品</u>，保证收纸完美。

(a) 测量最后一张印张色彩并将其提交给裁判。

(b) 从收纸堆顶部向下数第 30 张印张的套准和色彩将被评分。

8. <u>不要从印刷机上搬出收纸堆</u>，也不要将飞达上未使用的纸张取出。

9. 为下一项工作（顺利进行）清洁印刷机——<u>清洁辊</u>、橡皮布，清洁辊刮墨刀。清洁并整理工作区域。

请不要清除墨斗里的油墨。

10. 当你完成任务时示意裁判。

以上施工单选自第 45 届世界技能大赛印刷媒体技术项目的任务 1.3。

Chapter 2 Digital Printing

Unit 5 Tent Cards

Description

This job is Printed Digital Variable Data — competitors will receive a digital file via desktop and load it into digital RIP/server. Competitors will perform calibration procedure. Following instructions on job ticket, they will load drawer(s) with correct stock, merge data fields with 4/0 image to print variable data tent cards.

Job	**Variable Data Tent Cards**
Number of copies	52 tent cards
Set-up waste	10
Colors	4/0 – CMYK
Paper size	Sheet size 297 mm × 420 mm (A3)
Finished size	297 mm × 210 mm (A4)
Paper	200 g/m² Satin Digital
Ink	CMYK
Machinery (Digital Printing Press)	Heidelberg Versafire CV
Job status	Digital Files on desktop of Prinect RIP and on Preflight computer.
	15 minutes to preflight file.
	45 minutes to print – **no overtime will be allowed.**
Available time for the total task	This includes calibration, registration, loading digital files, creating layout, sizing the layout to fit the paper, setting paper drawers, completing quantity.

Before the printing	**Preflight** the file using a computer off-line from the press. Open the "WS_TentCard_WS_outline_Type_specs_1.pdf" file and the "data_tent_cards" file listing the number and names to be imprinted on the tent cards. Take note of the font, size and position information. Check data to be imported and organize a plan. **Production** on Prinect work station, select the "WS_TentCard_WS_outline" file and use the data file to format the files for variable data printing, on A3 paper. Print job information, bleed trim marks, 100% size. Perform calibration for the appropriate paper. You will be given 62 sheets of 200 g/m² satin digital paper. Competitors will load specified paper into press and set up.
During the printing	Monitor quality — verify the correct names and all names being printed are accounted for — have technicians fix machine jams if they occur.
After printing	Give required quantity to the judges.
Additional knowledge	This is a live job — tent cards printed will be used during meetings. Save cards to be trimmed in Task 12.

The job ticket above comes from task 2.1 of the Print Media Technology project of the 44th WSC.

▶ [VOCABULARY & PHRASE]

digital / ˈdɪdʒɪt(ə)l /
adj. 数字的
digital printing 数字印刷
digital file 数字文件
digital printing press 数字印刷机

desktop / ˈdesktɑːp /
n.（电脑）桌面；台式机
clear the desktop 清理桌面

server / ˈsɜːrvər /
n.（电脑）服务器
virtual server 虚拟服务器

calibration / ˌkælɪˈbreɪʃn /
n.（色彩）校正；（测量器具上的）刻度
calibration procedure （色彩）校正流程
screen calibration 屏幕校准

drawer / drɔːr /
n. 纸箱；抽屉
open the drawer 打开抽屉

stock / stɑːk /
n. 纸张；（商店的）库存；储备物；储备量；股本；股票；家畜
v.（商店或工厂）储备；为……备货
adj. 老一套的；（商店里）库存的
stock market 股市
stock exchange 证券交易所

merge / mɜːrdʒ /
v.（使）合并；融入；兼并（产权，产业）
merge together 混合起来

preflight / priːˈflaɪt /
adj. 预飞；起飞前的
preflight checks 飞行前检查

layout / ˈleɪaʊt /
n. 拼版；设计；版面设计
create layout 进行拼版
page layout 页面布局

imprint / ɪmˈprɪnt /
v. 印刷
n. 印记；痕迹
raindrop imprint 雨痕

font / fɑːnt /
n. 字体；根源，来源
font size 字体大小，字号
embed font 内嵌字体

import / ˈimpɔːrt /
v. 输入；导入（计算机）
n. 进口；输入；重要性
import duty 进口税
import licence 进口许口证

format / ˈfɔːrmæt /
n.（书或杂志的）版式
v. 为……编排格式；格式化
file format 文件格式
output format 输出格式

bleed / bliːd /
n. 出血，流血；泄出（液体，气体）
v. 出血，流血；给……放血；散开
bleed out 渗出
air bleed 排气阀

variable / ˈveriəb(ə)l /
adj. 可变的；易变的；时好时坏的
n. 可变性，可变因素
variable data 可变数据
control variable 控制变量

trim / trɪm /
v. 裁切；修剪；削减；修饰（尤指某物的边缘）
n.（尤指毛发的）修剪；额外装饰
adj. 整齐的；修长的
trim marks 裁切标记

fix / fɪks /
v. 维修；安排；处理（问题等）；固定；盯住；（注意力）集中在
n.（尤指简单、暂时的）解决方法
fix machine jams 排除设备故障

raster / ˈræstər /
n. [电子]光栅；试映图
Raster Image Processor（RIP）光栅图像处理器

Prinect 印通
job ticket 施工单
take note 做笔记
work station 工作站
job information 工作信息
monitor quality 监控质量
live job 真实的工作

▶ [TRANSLATION]

第 5 单元　席卡

描述

这个工作是数字印刷可变数据。选手通过（电脑）桌面获得数字文件，并将数字文件载入 RIP/ 服务器。选手需执行（色彩）校正流程。根据施工单说明，选手将正确的纸张装入纸箱，并将数据文件与 4/0 图像合并，以印刷可变数据席卡。

工作	可变数据席卡
印品数量	52 张席卡
校版纸	10
色彩	4/0 – CMYK
纸张规格	297 mm×420 mm (A3)
成品规格	297 mm×210 mm (A4)
纸张	200 g/m² 数字印刷光面纸
油墨	CMYK
机器（数字印刷机）	海德堡凌图 CV
工作状态	数字文件在印通 RIP 电脑桌面和预飞电脑上。
工作总时间	15 分钟用于预飞文件。 45 分钟用于印刷，**不允许超时。** 这包括（色彩）校正、套准、加载数字文件、拼版、调整版面以适应纸张尺寸、设置纸箱、印刷所需数量印品。

印刷前	使用与印刷机离线的电脑进行文件**预飞**。打开"WS_TentCard_WS_outline_Type_specs_1.pdf"和"data_tent_cards"文件，文件中列出了印刷在席卡上的编号和姓名。 记录（可变数据的）字体、尺寸和位置信息。 查看需要输出的数据并制订（工作）计划。 在印通工作站上进行**生产**，选择"WS_TentCard_WS_outline"文件，使用数据文件进行可变数据印刷，版式符合 A3 纸张。 印刷工作信息，出血裁切标记，无缩放。 使用合适的纸张进行（色彩）校正。 你将获得 62 张 200 g/m² 的数字印刷光面纸。 选手将规定的纸张装入印刷机并进行设置。
印刷中	质量监控是为了核实所有姓名及数量是否正确印刷，且在出现设备故障时由技术人员进行排除。
印刷后	将所需数量的（印品）交给裁判。
附加知识	这是一个真实的工作，印刷出的席卡将在会议中被使用。 保留席卡，席卡在任务 12 中会被裁切。

以上施工单选自第 44 届世界技能大赛印刷媒体技术项目的任务 2.1。

Part 1 Job Ticket

Unit 6 *Vision 2025* Booklet

Description

This job is Printed Digital — competitors will receive a digital file via desktop and load it into digital RIP/server. Competitors will perform calibration procedure.

Job	**Saddle Stitched *Vision 2025* Booklet (20 pages + cover)**
Number of copies	75
Set-up waste	5 sheets cover, 25 sheets content
Colors	4/4 Cover 4/4 Text
Paper size	Sheet size SRA3 320 mm × 450 mm
Finished size before cutting	SRA4 320 mm × 225 mm
Finished size after cutting	A4 210 mm × 297 mm
Paper	150 g/m² Satin Digital Cover 100 g/m² Satin Digital
Ink	CMYK
Machinery (Digital Printing Press)	Heidelberg Versafire CV
Job status	Digital files on desktop of Prinect RIP and preflight computer.
Available time for the total task	15 minutes to prepare the preflight file. 45 minutes to print — **no overtime will be allowed**. This includes calibration, registration, loading digital files, creating layout, setting paper drawers, completing quantity.

Before the printing	You will print a 20-page content with a 4-page cover. **Preflight** the file using a computer off-line from the press. Open the "Vision 2025" booklet file and preflight the file. Take notes of page size, trim size and bleeds. **Production — NO** calibration needed. You will need to perform a front to back registration test and adjust accordingly. The booklet will print 4/4 on the cover and 4/4 on the text — ensure that you have 3 mm bleeds. Create job label in Prinect RIP – "WSI Vision 2025 24p and your country code". Competitors will load specified papers into press and set up. They shall produce 75 finished copies from supplied PDF, ensuring that the image is correct on sheet. Upon completion of press run, give your produced saddle stitched booklets to the judges.
During the printing	Monitor quality — ask technician to fix machine jams if required.
After printing	Give required quantity to judges.
Additional knowledge	Finishing is an additional task 14. This includes preparing a cutting plan and cutting the printed product to the required dimensions in the cutting task 14.

The job ticket above comes from task 2.3 of the Print Media Technology project of the 44th WSC.

[VOCABULARY & PHRASE]

booklet / ˈbʊklət /
n. 小册子
residence booklet 户口簿

page / peɪdʒ /
n.（书、报纸、文件等的）页，面
page size 页面尺寸

cover / ˈkʌvər /
n. 封面；盖子；掩护；被子
v. 覆盖；涉及；报道；翻唱；行走（一段路程）
hard cover 精装
soft cover 平装

text / tekst /
n.（书、杂志等中区别于图片的）正文，文字；文字材料
v.（用手机）给……发短信
text editor 文本编辑器
text style 字体样式

cutting / ˈkʌtɪŋ /
n. 裁切；剪报
cutting plan 裁切计划

finishing / ˈfɪnɪʃɪŋ /
n. 印后加工；得分的表现和技巧
v. 完成；结束；用完，吃光（finish 的现在分词）
surface finishing 表面修整

dimension / daɪˈmenʃn; dɪˈmenʃn /
n. 尺寸；（空间的）维度；规模；方面
v. 切削（或制作）成特定尺寸；标出尺寸
three dimension 三维

saddle / ˈsæd(ə)l /
n. 鞍，马鞍
v. 使负重担；跨上马鞍
saddle stitching 骑马订
bicycle saddle 自行车座

label / ˈleɪb(ə)l /
n. 标签；称号；商标；唱片公司；（计算机）标记
v. 贴标签；把……不公正地称为
job label 工作标签
logo label 商标

additional / əˈdɪʃən(ə)l /
adj. 附加的，额外的
additional tax 附加税

trim size 裁切尺寸
country code 国家代码
finished copies 印品

▶ [TRANSLATION]

第 6 单元 《展望 2025》小册子

描述
这个工作是数字印刷。选手通过（电脑）桌面获得数字文件，并将数字文件载入 RIP/服务器。选手需执行（色彩）校正流程。

工作	骑马订《展望 2025》小册子（20 页内页＋封面）
印品数量	75 本
校版纸	5 张封面纸，25 张内页纸
色彩	4/4 封面 4/4 文字
纸张规格	SRA3 320 mm×450 mm
裁切前规格	SRA4 320 mm×225 mm
裁切后规格	A4 210 mm×297 mm
纸张	150 g/m² 的数字印刷光面封面纸 100 g/m² 的数字印刷光面纸
油墨	CMYK
机器（数字印刷机）	海德堡凌图 CV
工作状态	数字文件在印通 RIP 电脑桌面和预飞电脑上。
工作总时间	15 分钟用于准备预飞文件。 45 分钟用于印刷，**不允许超时**。 这包括（色彩）校正、套准、加载数字文件、拼版、设置纸箱、完成所需数量印品。

	你将印制一本 20 页内页和 4 页封面的小册子。
	使用与印刷机离线的电脑进行文件**预飞**。打开《展望 2025》小册子文件并进行预飞。
	记录页面尺寸、裁切尺寸和出血。
	生产，不需要进行（色彩）校正。
印刷前	你需要进行正反面套准测试和做相应的调节。
	小册子的封面和内页文本均为 4/4 印刷，确保有 3 mm 出血。
	在印通 RIP 里添加工作标签，标签内容为"WSI Vision 2025 24p 和你的国家代码"。
	选手将指定的纸张装入印刷机并进行设置。根据提供的 PDF 文件印刷 75 本印品，并确保纸张上的图像正确。
	印刷完成后，将你生产的骑马订小册子交给裁判。
印刷中	质量监控，即出现设备故障时呼叫技术人员进行排除。
印刷后	将所需数量的印品交给裁判。
附加知识	印后加工为附加任务 14。这包括在裁切工作 14 中制订裁切计划，并将印刷好的产品裁切至所需尺寸。

以上施工单选自第 44 届世界技能大赛印刷媒体技术项目的任务 2.3。

Unit 7 Preflight and Edit Digital Image

Description

The digital press operators must be able to open and edit PDF files to make changes required by the customers.

Marking

Measurement assessment will be used.

Time allotment

20 minutes

Task

This task will require the use of Adobe Acrobat Software — Using Adobe Acrobat, competitors will open a PDF file of a business card template provided by the judges. The competitors are to preflight the file to assure printability and edit the copy as directed.

1. When the judge says to begin, open the provided PDF files with Acrobat.

2. Preflight the file to assure it contains bleed trim images and the image is printable.

3. Edit the PDF with Acrobat so that the business card contains your personal information:

(a) Replace the provided fields with your name, phone, email.

(b) Change the title to "WorldSkills Competitors".

(c) Make sure the font, style, and size do not change.

4. Save the file as "your country_business card", for example: MX_business card.pdf.

5. Notify the judges when you are completing.

The job ticket above comes from task 2.1 of the Print Media Technology project of the 45th WSC.

▶ [VOCABULARY & PHRASE]

edit / ˈedɪt /
v. 编辑，校订；编选；剪辑；主编
n. 编辑，校订；剪辑
text edit 文本编辑器

template / ˈtemplət /
n.（计算机文档的）模板；样板；垫木
template parameter 模板参数
web template 网页模板

customer / ˈkʌstəmər /
n. 顾客
customer service 客户服务

printability / ˌprɪntəˈbɪlɪtɪ /
n. 印刷适性，可印染的
printability test 印刷适性测试

software / ˈsɔːftwer /
n. 软件
software development 软件开发
software license 软件许可证

style / staɪl /
n.（字）形；方式；款式
v. 设计，给……造型；称呼，命名

▶ [TRANSLATION]

第 7 单元　预飞和编辑数字图像

描述

数字印刷机操作员必须能够打开和编辑 PDF 文件，并按照客户要求对 PDF 文件进行更改。

评分

将使用测量评分进行评判。

时间分配

20 分钟

任务

这个任务需要使用 Adobe Acrobat 软件。使用 Adobe Acrobat，选手打开裁判提供的 PDF 名片模板。选手对文件进行预飞，以保证文件的印刷适性，并按指令编辑副本。

1. 当裁判宣布开始后，使用 Acrobat 打开提供的 PDF 文件。

2. 对文件进行预飞，确保文件包含出血图像，以及图像可印刷。

3. 使用 Acrobat 编辑 PDF 文件，使名片中包含你的个人信息：

(a) 将提供的文件改为你的姓名、电话和邮箱。

(b) 将标题改为"WorldSkills Competitors"。

(c) 确保字体、字形、字号无变化。

4. 以"你的国家_business card"为名称保存文件，例如：MX_business card.pdf。

5. 当你完成任务时示意裁判。

以上施工单选自第 45 届世界技能大赛印刷媒体技术项目的任务 2.1。

Unit 8　Preflight

Description

The digital press operators must be able to preflight PDF files to assure the best quality will be printed for the customers. Many times, customers will provide digital images not knowing they are not correct resolutions.

Marking

Measurement assessment will be used.

Time allotment

20 minutes

Task

This task will require the use of Adobe Acrobat software.

1. When the judge says "begin", open the digital file provided "Post Cards for Preflight" with Adobe Acrobat.

2. Preflight the digital file to check following specifications for the usual faults.

(a) Bleeds: Over 2 mm (image extends beyond the trim line).

(b) Image resolution: Over 200 PPI.

(c) PDF Standard: PDF/X-4:2010.

(d) PDF Output Intent Profile: Coated FOGRA39.

(e) Colors: CMYK + Pantone 2955C.

3. Preflight the digital files to determine which files are of acceptable specification for the printing of postal cards.

4. Complete the form on next page and define if the file is acceptable. You should mark (a, b, c, d or e) for the reasons of incorrect file.

5. Create a single new file with only the files that are acceptable specifications.

Name the file "your country_postcards" (like *MX_postcards*).

6. Notify the judges when you have done.

File Name	Correct = Yes Incorrect = No	If Incorrect, mark the fault /reasons for incorrect file: a. Bleeds b. Resolution size c. PDF Standard d. PDF Output Intent Profile e. Colors
Example WSC2019_2.1.N	No	B, D
Example WSC2019_2.1.M	Yes	
WSC2019_Task_6_A		
WSC2019_Task_6_B		
WSC2019_Task_6_C		
WSC2019_Task_6_D		
WSC2019_Task_6_E		
WSC2019_Task_6_F		
WSC2019_Task_6_G		
WSC2019_Task_6_H		

The job ticket above comes from task 2.3 of the Print Media Technology project of the 45th WSC.

[VOCABULARY & PHRASE]

resolution / ˌrezəˈluːʃn /
n.（电视、照相机、显微镜等的）分辨率，清晰度；正式决定；（冲突、问题等的）解决办法；决心
image resolution 图像分辨率
high resolution 高分辨率
dispute resolution （法律）调解纠纷
conflict resolution 冲突解决
low resolution 低分辨率

specification / ˌspesɪfɪˈkeɪʃ(ə)n /
n. 规格；规范；说明书；详述
technical specification 技术规范
design specification 设计规格

fault / fɔːlt /
n. 故障；错误
v. 挑剔，指责
page fault 页面错误

fault finding 排错

coated / ˈkotɪd /
adj. 涂布的；覆盖着的
v. 外面覆盖（coat 的过去分词）
coated paper 刮刀涂布纸

Pantone / pænton /
n. 潘通色卡
PANTONE FORMULA GUIDE 彩通配方指南

intent / ɪnˈtent /
n. 目的；故意
adj. 专注的；坚决的；急切的
output intent profile 输出目标特征文件

trim line 裁切线

[TRANSLATION]

第 8 单元　预飞

描述

数字印刷机操作员必须能够对 PDF 文件进行预飞，以保证给客户的印刷品质量最佳。很多时候，客户自己也不知道他们提供的数字图像分辨率不对。

评分

将使用测量评分进行评判。

时间分配

20 分钟

任务

这个任务需要使用 Adobe Acrobat 软件。

1. 当裁判宣布开始后，使用 Adobe Acrobat 打开提供的"Post Cards for Preflight"数字文件。

2. 按照下述规范对数字文件进行预飞，以查看是否存在常见错误。

(a) 出血：大于 2 mm（图像超出裁切线的部分）。

(b) 图像分辨率：大于 200PPI。

(c) PDF 标准：PDF/X-4:2010。

(d) PDF 输出目标特征文件：Coated FOGRA39。

(e) 色彩：CMYK + Pantone 2955C。

3. 对数字文件进行预飞，判断哪些文件符合明信片印刷可接受的规格。

4. 根据文件是否被接受，填写下一页的表格。错误文件的原因需标注（a、b、c、d 或 e）。

5. 将符合规格的文件合并成一个新文件。

将该文件以"你的国家_postcards"为名称命名（例如 MX_postcards）。

6. 当你完成任务时示意裁判。

Part 1 Job Ticket

文件名	正确 = Yes 错误 = No	若错误，填写错误文件的原因： a. 出血 b. 分辨率大小 c. PDF 标准 d. PDF 输出目标特征文件 e. 色彩
例子 WSC2019_2.1.N	No	B、D
例子 WSC2019_2.1.M	Yes	
WSC2019_Task_6_A		
WSC2019_Task_6_B		
WSC2019_Task_6_C		
WSC2019_Task_6_D		
WSC2019_Task_6_E		
WSC2019_Task_6_F		
WSC2019_Task_6_G		
WSC2019_Task_6_H		

以上施工单选自第 45 届世界技能大赛印刷媒体技术项目的任务 2.3。

Unit 9 Engineer Workflow and Print a Digital File

Description

The digital press operators must be able to maximize the efficiency of the printing company by building a "sequence folder" or "hot folder" workflow that can be used for this and future jobs.

Marking

Judgment and measurement assessment will be used.

Time allotment

30 minutes

Task

This task will require the use of the Prinect RIP software and the operation of the Versafire digital press.

You should work like normal print house. Environmentally friendly and keep workspace clean.

1. When the judge says "begin", open the "Postcards for Preflight" PDF file you created in the preflight exercise or a file provided to you by the judge.

2. Using the Prinect RIP, create a page list or hot folder that will produce multiple post cards on a single press sheet.

3. The page list/hot folder must include the following:

(a) Use a 4-up imposition.

(b) Automatic selection of 200 gsm Satin as the paper to be printed.

(c) The placement of trim and registration marks on both sides.

(d) Automatic printing of 10 press sheets when the print button has been pressed.

(e) Register the front to back images to within 0.6 mm in each corner of the press sheet.

4. Make front / back registration and give the judges each of your incorrect registration sheets.

5. Give the judge your "OK" sheet press sheet.

6. Give all make ready sheets to the judges.

7. <u>Print 10 copies</u> of your post cards.

8. Notify the judges when you have completed.

The job ticket above comes from task 2.4 of the Print Media Technology project of the 45th WSC.

▶ [VOCABULARY & PHRASE]

efficiency / ɪˈfɪʃ(ə)nsi /
n. 效率；（机器的）功率
economic efficiency 经济效率
high efficiency 高效率

workspace / ˈwɜːkspeɪs /
n. 工作区域
close workspace 关闭工作区

workflow / ˈwɜːkfloʊ /
n. 工作流程
engineer workflow 工程工作流程

sequence / ˈsiːkwəns /
n. 顺序；连续事件（或动作、事物）
v. 按顺序排列；测定（整套基因或分子成分的）序列；用音序器播放（或录制）音乐

sequence folder 序列文件夹
hot folder 热文件夹
print house 印刷厂
environmentally friendly 环保
page list 页面列表

▶ [TRANSLATION]

第9单元　工程工作流程和印刷数字文件

描述

　　数字印刷机操作员必须能够为当下或未来的工作建立一个"序列文件夹"或"热文件夹"工作流程，以此最大限度地提高印刷企业的效率。

评分

将使用评价评分和测量评分进行评判。

时间分配

30 分钟

任务

本任务需要使用印通 RIP 软件并操作 Versafire 数字印刷机。

你应该像平时在印刷厂进行工作一样。坚持环境保护和保持工作区域清洁。

1. 当裁判宣布开始后，打开你在预飞任务中创建的"Postcards for Preflight" PDF 文件或裁判提供给你的文件。

2. 使用印通 RIP，创建一个页面列表或热文件夹，这个页面列表或热文件夹可支持将多种明信片印刷在一张纸上。

3. 页面列表或热文件夹需包含以下内容：

(a) 使用 4-up 拼版。

(b) 印刷时自动选择 200 g/m² 的光面纸。

(c) 印张双面均放置裁切和套准标记。

(d) 当点击印刷按钮时，自动印刷 10 张纸。

(e) 页面四角图文正反面套准需在 0.6 mm 以内。

4. 进行正/反面套准，将未套准的印张交给裁判。

5. 将你的"OK 样"交给裁判。

6. 将所有的校版纸交给裁判。

7. 印刷 10 份明信片。

8. 当你完成任务时示意裁判。

以上施工单选自第 45 届世界技能大赛印刷媒体技术项目的任务 2.4。

Part 1 Job Ticket

Unit 10 Color Correct and Print a Digital File

Description

The digital press operators must be able to make color corrections to digital files to match products the customer had previously printed elsewhere.

Marking

Judgment and measurement assessment will be used.

Time allotment

30 minutes

Task

This task will require the use of the Prinnect RIP software and the operation of the Versafire digital press.

When the judge says "begin", open the "WSCT-Postcard_print. pdf".

1. Use the Prinnect RIP and Versafile Press to print a copy of the file on 200 gsm Satin paper.

2. Use the color management tools to:

(a) Read the Lab values of samples the customer provided and the Versafire printed samples.

   (i) There is one area marked for you to read.

(b) Make adjustments to the Versafire color print to best match the samples the customer provided.

3. Give the judges all of your printed sheets used to adjust color.

4. Give the judges your final "OK" sheet.

5. Give all make ready sheets to the judges.

6. <u>Print 10 copies</u> of your color corrected job.

7. Notify the judges when you have completed.

The job ticket above comes from task 2.6 of the Print Media Technology project of the 45th WSC.

▶ [VOCABULARY & PHRASE]

adjust / əˈdʒʌst /

v. 调整；整理（衣服）；习惯；评估（损失，损害）

adjust to 适应

color correct 色彩修正

color management 色彩管理

▶ [TRANSLATION]

第 10 单元　色彩修正和印刷数字文件

描述

数字印刷机操作员必须能够对数字文件进行色彩修正，使该色彩与客户之前在其他地方印刷的产品的色彩相匹配。

评分

将使用评价评分和测量评分进行评判。

时间分配

30 分钟

任务

本任务需要使用印通 RIP 软件并操作 Versafire 数字印刷机。

当裁判宣布开始后，打开"WSCT-Postcard_print.pdf"文件。

1. 使用印通 RIP 和 Versafile 印刷机，在 200 g/m² 的光面纸上印刷一份文件。

2. 使用色彩管理工具完成下列任务：

(a) 读取客户提供样张和 Versafire 印刷样张的 Lab 值。

　　(i) 在标注的区域进行（数据）读取。

(b) 对 Versafire 印刷色彩进行调整，使其与客户提供的样张完美匹配。

3. 将你进行色彩调整的印张全部交给裁判。

4. 将你的最终"OK 样"交给裁判。

5. 将校版纸全部交给裁判。

6. 印刷 10 张色彩正确的印张。

7. 当你完成任务时示意裁判。

以上施工单选自第 45 届世界技能大赛印刷媒体技术项目的任务 2.6。

Chapter 3 Additional Task

Unit 11 Maintenance

Description

A major responsibility of the offset press operators is to safely install and adjust rollers, plates, blankets on the offset printing press. There is a procedure for preforming these tasks to assure safety to the operators and to the equipment. Competitors will be evaluated on speed and the accuracy of their settings and the process by which they perform the tasks.

Number of copies	Not Applicable
Machinery	Heidelberg Speedmaster SM-52-2
	Roller Stripe Gauge
	Micrometer
	Pen and paper
	Plate
Job status	The press will not have a blanket and #4 form roller installed.
Available time for the total task	One hour — **no overtime will be allowed**
Tasks to be done	1. Build the blanket and packing to a thickness of 3 mm for installation on press and choose the right prepared packages most suitable.
	2. Mount the blanket with packing on press.
	3. Install #4 ink form roller into press.
	4. Insert the plate. Set roller stripe on all four ink form rollers and dampening roller according to the instructions (there is ink on the rollers).

Part 1 Job Ticket

Tasks to be done

INSTRUCTION - All stripes are to be 3 mm (ink rollers) and 5 mm (dampening roller).

Begin when the experts tell you to start.

When you have completed the task, notify the experts to stop the clock.

Speed is relevant to this task according to the quality.

The job ticket above comes from task 3.1 of the Print Media Technology project of the 44th WSC.

▶ [VOCABULARY & PHRASE]

maintenance / ˈmeɪntənəns /
n. 维护，保养；（依法应承担的）抚养费
maintenance area 维修区

equipment / ɪˈkwɪpmənt /
n. 设备；（做某事应具备的）素质，才能
peripheral equipment 外部设备
medical equipment 医疗器械

micrometer / maɪˈkrɑːmɪtər /
n. 千分尺
depth micrometer 深度千分尺

packing / ˈpækɪŋ /
n. （橡皮布）包衬；包装；填充物
v. 包装
packing list 包装单

package / ˈpækɪdʒ /
n. （橡皮布）包衬；包裹；（建议或提供的想法或服务的）一套
v. 把……打包；包装（产品或想法等）
package size 包装尺寸

thickness / ˈθɪknəs /
n. 厚度，粗细；浓度；密度；最厚（或最深）处
v. 刨，削（木头至合适的尺寸）
film thickness 漆膜厚度

tasks to be done 必要工作
dampening roller 水辊
additional task 附加任务
safely install 安全安装
roller stripe gauge 辊压痕量规
ink form roller 着墨辊

▶ [TRANSLATION]

第 11 单元　维护保养

描述

胶印机操作员的主要职责之一是在胶印机上安全安装并调节辊、印版和橡皮布。对操作员来说，为确保自身和设备安全，有一套执行这些工作的流程。选手在执行工作时的速度、设置准确性和过程将被评估。

印张数量	不适用
	海德堡速霸 SM-52-2
	辊压痕量规
机器	千分尺
	笔和纸
	印版
工作状态	印刷机上的橡皮布和 4 号着墨辊未安装。
工作总时间	1 小时，**不允许超时**。
	1. 选择最合适的包衬，使橡皮布和包衬构成 3 mm 的厚度，以便在印刷机上安装。
	2. 在印刷机上安装橡皮布和包衬。
	3. 在印刷机上安装 4 号墨辊。
必要工作	4. 安装印版。根据说明调节所有墨辊和水辊（辊上已经有油墨）。
	说明：所有压痕为 3 mm（墨辊）和 5 mm（水辊）。
	专家宣布开始后才可工作。
	当你工作完成后示意专家停止计时。
	根据完成质量，完成速度也和本工作相关。

以上施工单选自第 44 届世界技能大赛印刷媒体技术项目的任务 3.1。

Part 1 Job Ticket

Unit 12 Defect Detection

Description

This is a timed event. When you are ready to begin, you will notify the judge to start the clock. When the clock starts, you may open your packet and begin to inspect each of the (9) press sheets. You are to <u>circle all defects</u> you find on each of the press sheets and <u>indicate on the marking sheet what type, and how many defects</u> you find on each of the sheets. When you finish, notify the judges to stop the clock.

Job	**Identify common offset printed defects**
Number of copies	each of 9 different defective press sheets
Machinery	Loop magnifying glass and permanent marker
Job status	Complete
Available time for the total task	30 minutes — **no overtime will be allowed.** Competitors will be timed. The competitors with the fastest time will receive full time marks. Competitors with the slowest time will receive 0 time marks. Each competitor in between the fastest and the slowest will receive their fraction of the time mark.
During this task	When you are ready to begin, notify the judges and begin when the judges tell you to start. Open your packet. Review each of the nine press sheets. Using the permanent marker, <u>circle each of the defects</u> you find on each of the sheets. Locate the corresponding box for each sheet and record the <u>number of defects</u>, the <u>classification</u> of defect found for that sheet, and what <u>color of ink or plate</u> that the defect was located.
After this task	When you have completed the task, notify the judges to stop the clock. Place your sheets and marking sheet back into the packet and give them to the judge for marking.

Each of the defective sheets in the packet will have a minimum of 1 defect and a maximum of 4 defects.

The defects that may or may not be on the sheets are limited to these types:

Front-to-Back registration is when the image on side A of the press sheet does not align with side B or the image on one side of the sheet is not correctly oriented for when the printed sheet is finished.

Hickeys are caused by dirt, paper fibers, or hardened specks of ink on the printing plate or blanket.

Incorrect color of ink is when the wrong plates are placed on the wrong printing unit or the plates have been installed on the wrong unit.

Additional knowledge

Incorrect placement of paper in the feeder is when there is a specific characteristic of the paper that requires that it must be placed correctly into the feeder. Such as a coated one sided sheet, or one of several other instances.

Registration Defects are when one or more of the colors being printed do not align properly.

Scratches are caused when something rubs against the plate at any time, it may scratch the surface making it ink receptive. Those scratches will print on the paper.

Scumming appears when a non-image area such a reverse type begins to fill in with color due to poor ink and water balance.

Smashed Blanket will show an area of missing image by not transfering ink to the surface of the paper.

The job ticket above comes from task 3.2 of the Print Media Technology project of the 44th WSC.

[VOCABULARY & PHRASE]

orient / ˈɔːrɪent /
v. 朝向；确定方位；引导；使熟悉
n.（旧）东方，东亚诸国（the Orient）；优质珍珠
adj.（文）东方的；（尤指宝石）光彩夺目的；（太阳等）冉冉上升的
Orient Express 东方快车
Pearl of the Orient 东方之珠

scratch / skrætʃ /
n.（某人皮肤上的）划痕；抓，挠
v.（用指甲）挠，轻抓；划出（痕迹）；（用爪子）刨；勾掉（写下的字）
adj.（球队或一群人等）仓促拼凑的，匆匆组成的
start from scratch 从零开始
cat-scratch disease 猫抓病

detection / dɪˈtekʃ(ə)n /
n. 鉴别；察觉；侦破（案件）
defect detection 弊病鉴别

loop / luːp /
n. 环形；（程序中）一套重复的指令；回路；〈英〉（铁道或公路的）环线
v. 使绕成圈；执行计算机指令
event loop 事件循环

magnify / ˈmæɡnɪfaɪ /
v. 放大；夸张；使（问题）加重
loop magnifying glass 环形放大镜

permanent / ˈpɜːmənənt /
adj. 永久的；（尤指问题或困难）一直存在的
permanent marker 记号笔
permanent resident 永久居民

align / əˈlaɪn /
v. 使一致；公开支持；（使）排成一条直线
align with 重合
text-align 文本对齐

hickey / ˈhɪkɪ /
n. 墨皮；器械；唇印

fiber / ˈfaɪbər /
n. 纤维（等于 fibre）
paper fibers 纸张纤维
glass fiber 玻璃纤维

balance / ˈbæləns /
n. 天平；平衡；账户余额
v. 保持平衡
ink and water balance 水墨平衡

smash / smæʃ /
v. 撞毁（交通工具）；打碎；（使）猛撞；（轻松）打破（纪录）
n. 猛烈撞击声；猛击
adv. 哗啦一声
adj. 非常轰动的，出色的
smashed blanket 橡皮布塌陷
smash shot 扣杀球

speck / spek /
n. 污点；灰尘；小颗粒
v. 使有斑点
paper speck 纸斑

receptive / rɪˈseptɪv /
adj. 可以接受的；（对观点、建议等）愿意倾听的
receptive unit 感觉单位

scumming / ˈskʌmɪŋ /
n. 起脏；撒渣；吐渣
v. 除去（浮渣）
de-scumming 去渣

reverse / rɪˈvɜːrs /
adj. 相反的；反向的
v. 逆转（决定、政策、趋势等）；撤销（法庭判决）；颠倒；交换（位置、功能）
n. 逆向，逆转
reverse psychology 逆反心理

specific / spəˈsɪfɪk /
adj. 特定的；具体的；有特效的
n. 细节；特效药
specific instructions 明确的指示

transfer / trænsˈfɜːr /
v. 转印（图画，图案）；（使）转移；传染（疾病）；转让（权力等）
n. 转移，调动；（旅行中）转乘；〈美〉换乘票；纸上可转印的图画或图案
technology transfer 技术转移
wire transfer 电汇

timed event 计时工作
marking sheet 答题纸
common offset printed defects 常见胶印弊病
front-to-back registration 正反面套准
specific characteristic 特定特征
coated one sided 单面涂布
non-image area 非图像区域

▶ **[TRANSLATION]**

第 12 单元　弊病鉴别

描述

这是一项按完成时间排序的工作。当你准备好后示意裁判开始计时。计时开始后，打开袋子开始查看每一张印张（共 9 张）。你需要圈出在每张印张上找到的所有弊病，并在答题纸上写出每张印张的（弊病）类型和弊病数量。当你完成工作后示意裁判停止计时。

工作	常见胶印弊病鉴别
印品数量	9 张不同弊病印刷样张的每 1 张
机器	环形放大镜和记号笔
工作状态	完成（工作）
总工作时间	30 分钟，**不允许超时**。
	选手工作将被计时。最快完成的选手时间分为满分。最慢完成的选手时间分为 0 分。最快和最慢之间的选手（按时间从快到慢）分段得分。
工作中	当你准备好开始时，示意裁判，并在裁判宣布开始后开始工作。打开你的袋子，查看 9 张印张的每 1 张。使用记号笔，<u>圈出</u>你在每张印张中找到的<u>每个弊病</u>。在每张印张对应的框中记录<u>弊病数量</u>，找到的弊病<u>类型</u>，以及弊病出现在哪个<u>颜色</u>或（哪个颜色的）<u>印版上</u>。
工作后	当你完成工作后，示意裁判停止计时。将样张和答题纸装回袋子，并将袋子交给裁判进行打分。

袋子中的每张有弊病的样张至少有 1 个弊病，最多有 4 个弊病。

样张上有或没有的弊病仅限于以下类型：

附加知识

正反面套准（弊病） 出现在印张 A 面的图像和 B 面的图像不重叠或印张上一面的图像在完成印刷时方向错误的情况下。

墨皮（弊病） 是由印版或橡皮布上的的污物、纸张纤维或油墨杂质造成的。

色彩错误（弊病） 是指将错误的印版装在错误的印刷机组内或将印版装在错误的机组内。

飞达上的纸张放置错误（弊病） 是指有特定特征的纸张要求必须将其正确装入飞达。例如单面涂布纸，或其他的例子。

附加知识

套准错误（弊病） 是指在一个或多个颜色印刷时没有正确重叠。

划痕 有时是异物与印版摩擦造成的，印版表面被划后油墨会渗入划痕。这些划痕将被印刷到印张上。

起脏 出现在非图像区域被印上色彩时，而这又是水墨不平衡造成的。

橡皮布塌陷 表现为油墨未转移至纸张表面而造成的区域图像缺失。

以上施工单选自第 44 届世界技能大赛印刷媒体技术项目的任务 3.2。

Part 1 Job Ticket

Unit 13 Fault finding, maintenance

Description

The world class press operators must be able to see small defects, be able to identify their causes, correct them and print a superior product in a short amount of time.

Marking

Judgment and measurement assessment will be used.

Time allotment

2 hours total — 30 minutes for maintenance and fault finding, 1 hour and 30 minutes for printing mixed ink.

Task

30 minutes

This task will require the competitors to work in an isolated workspace with Heidelberg technician and master printer Stephan Boes. Stephan's job is to observe you as you work to solve the problems of this task. He is there to assure your safety and the safety of the equipment. He will also keep a record of the number of process steps, number of attempts, the correct use of measurement equipment, results of your measurements, and the time of this task.

He is **not** there to assist you in the completion of the task.

1. When the competitors are ready to begin, notify the judges to start the timing.

2. Install blanket on press.

(a) Select correct packing from assortment of packing papers.

(b) Follow proper procedures for blanket installation.

3. Make ready the feeder and delivery of the press with 1000 sheets of gloss adhesive stock. To complete this task, the press should be ready to add plates and ink.

4. Notify the judges when this task is completing.

Task

1 hour and 30 minutes

This task will require the competitors to print 500 sellable press sheets of stickers using the ink mixed in previous task.

1. Place the Pantone ink you mixed into the ink fountain designated by the technician.

Plates are on the press and correct — no faults here.

2. Register the image to the paper and color to color.

3. Inspect the sheet carefully for defects.

4. Bring your mixed ink color to match your reference sample.

(a) PANTONE 332c

(b) PANTONE 2617c

5. Measure the Lab values of your mixed ink.

(a) You will use the same sample swatch used to mix ink.

(b) If may, adjust your formula to bring closer to Lab target.

(c) Give the technician your final "OK" press sheet and indicate your spot for measurement by circling with template.

(d) Print 500 press sheets.

(e) Do not remove the finished sheets from the cart.

6. Clean the Press.

(a) **Remove the ink from the fountain**, wash the roller, clean the blanket and wipe the roller clean up blade.

(b) Wash rollers and ink fountain — do not remove wash-up attachment.

(c) Do not remove plates.

(d) Press and working area must be clean and all tools are put away.

7. **Remove the blanket and packing** from the same cylinder you found at the start of task.

Notify the judges when you are completing.

The job ticket above comes from task 3.2 of the Print Media Technology project of the 45th WSC.

▶ [VOCABULARY & PHRASE]

sticker / ˈstɪkər /
n. 不干胶；（有图或文字的）粘贴标签，贴纸
bar coded sticker 条形码贴纸

match / mætʃ /
v. 匹配；使相配；与……一致；使较量；适应
n. 比赛；火柴；相配的人（或物）；配偶；匹配
perfect match 完美组合
The Little Match Girl 卖火柴的小女孩

cart / kɑ:rt /
n. 收纸台板；马车；小型机动车；购物车，手推车
v. 用车装运；搬运；挟持，带走
hand cart 手推车

tool / tu:l /
n. （尤指手用）工具；蠢人；（书籍装订时的）压印图案
v. 驱车兜风；（用工具）制作，（在皮革，尤指书籍的皮革封面上）压印图案
power tool 电动工具

isolate / ˈaɪsəleɪt /
v. 隔离；分离；区别看待
adj. 孤独的，孤立的
n. 被隔离的人（或物）
isolated workspace 隔离的工作区域

attempt / əˈtempt /
v. 尝试，努力
n. 试图；企图杀害；（运动员创造纪录的）尝试，冲击
attempt to 尝试

adhesive / ədˈhi:sɪv /
adj. 黏着的，有黏性的
n. 黏合剂，胶水
gloss adhesive stock 高光不干胶

reference / ˈrefrəns /
adj. 参考的；文献索引的，参照的
n. 提及；参考；引文；参考书目；（帮助或意见的）征求；推荐信；介绍人
v. 提及；引用
reference sample 参考样张

attachment / əˈtætʃmənt /
n. （邮件）附件；连接物；附属物；依恋；（临时的）委派；信念；扣押
wash-up attachment 刮墨槽

swatch / swɑːtʃ /
n. 样本，样品
sample swatch 样张

fault finding 故障查找
mixed ink 调配好的（专色）油墨

process steps （工作）流程步骤
number of attempts 尝试（解决问题的）次数
measurement equipment 测量仪器
ink fountain 墨斗
roller clean up blade 辊清洁刮刀

[TRANSLATION]

第13单元　故障排除，维护保养

描述

拥有世界水平的印刷机操作员必须能够观察到小问题，能够确认问题成因，并能在短时间内解决问题，印刷出高质量产品。

评分

将使用评价评分和测量评分进行评判。

时间分配

共计2小时，30分钟用于维护保养和故障查找，1小时30分钟使用调配好的油墨来印刷。

任务

30分钟

本任务需要选手在一个隔离工作区域内与海德堡技术人员和印刷专家Stephan Boes共同完成。Stephan的工作是观察你在这项任务中解决问题（的过程）。他是为了确保你的人身安全和设备安全。他将记录你的（工作）流程步骤、尝试（解决问题的）次数、（是否）正确使用测量仪器、你的测量结果和工作用时。

他**不会**帮助你完成该任务。

1. 选手准备好后，示意裁判开始计时。

2. 在印刷机上安装橡皮布。

(a) 从多种包衬纸中选择正确的包衬。

(b) 按照恰当的流程安装橡皮布。

3. 使用1000张高光不干胶调节印刷机飞达和收纸。为完成这项任务，印刷机需保持能够安装印版并注入油墨的正常状态。

4. 完成该任务后示意裁判。

任务

1小时30分钟

本任务需要选手使用前序任务中调配好的（专色）油墨来印刷500张可销售的不干胶印张。

1. 将你调配好的潘通（专色）油墨注入技术人员指定的墨斗。

印刷机已经安装好印版，此处无误。

2. （调节）印张上图像的色彩与色彩套准。

3. 仔细查看印张上的弊病。

4. 将你调配好的（专色）油墨色彩与参考样张进行匹配。

(a) PANTONE 332c

(b) PANTONE 2617c

5. 测量你调配的（专色）油墨的Lab值。

(a) 你将使用和（专色）油墨调配时同样的墨样。

(b) 如果需要，（再次）调配你的（专色油墨）配方，使其更加接近Lab目标。

(c) 将你最终的"OK样"交给技术人员，在样张上圈出测量点并示意。

(d) 印刷500张印品。

(e) 不要将最终印品从收纸台板上搬走。

6. 清洁印刷机。

(a) **将油墨从墨斗中铲除**，清洁辊，清洁橡皮布并清洁辊清洁刮刀。

(b) 清洁辊和墨斗，不要取下刮墨槽。

(c) 不要拆除印版。

(d) 清洁印刷机和工作区域，并将工具复位。

7. 将你在本任务开始时（安装在）滚筒上的**橡皮布和包衬拆除**。

当你完成任务时示意裁判。

以上施工单选自第 45 届世界技能大赛印刷媒体技术项目的任务 3.2。

Part 1 Job Ticket

Unit 14 Ink Mixing

Description

Mix an appropriate quantity of Pantone ink, match reference, safety and cleanliness will be evaluated.

Job	**Mix Special Pantone Color**
Work Area	Work bench
Equipment available	Range of Pantone inks
	ink scale
	Pantone book
	calculator
	ink knives
	X-Rite eXact hand held
Available time for the total task	60 minutes to solve the task — **no overtime will be allowed**
Task to be done	Competitors will have to mix:
	450 grams of PMS 302
	Competitors will have to mix:
	450 grams of PMS 375
	You will measure each color of ink with the eXact hand held spectrodensitometer. When you have gotten the Lab you like, call the judges for final reading and provide a sample of your measured value to the judges.
	All mixed ink must be placed in container.
After solving the task	Upon completion of ink mixing, place product into supplied container.

	Work area must be left clean.
Additional knowledge	No ink waste allowed.
	Safety and cleanliness will be evaluated.

The job ticket above comes from task 4.1 of the Print Media Technology project of the 44th WSC.

▶ [VOCABULARY & PHRASE]

appropriate /əˈproʊprieɪt /	*n.* 计算器
adj. 合适的，相称的	pocket calculator 袖珍计算器
v. 拨出（款项）；私占，挪用	
appropriate penalty 适当的处分	spectrodensitometer /ˌspektrədensɪˈtɑːmɪtər /
be appropriate for 适合于	*n.* 分光光度仪
	double-beam spectrodensitometer 双光束分光光度仪
quantity / ˈkwɑːntəti /	
n. 量，数目	
quantity production 批量生产	container / kənˈteɪnər /
	n. 容器
safety / ˈseɪfti /	waste container 垃圾桶
n. 安全；安全性；安全场所	
adj.（特征、措施）保障安全的	bench / bentʃ /
safety goggle 安全眼罩	*n.* 长条形工作台；长凳；替补席
safety valves 安全阀	work bench 工作台
cleanliness / ˈklenlinəs /	scale / skeɪl /
n. 清洁	*n.* 磅秤；等级；刻度；规模；比例；鳞；音阶
surface cleanliness 表面洁净度	
	v. 改变（文字、图片）尺寸大小；刮去（鱼鳞）；攀登；剔除（牙垢）；称得重量为
calculator / ˈkælkjuleɪtər /	

adj.（模型或复制品）按比例缩小的

ink scale 油墨秤

economies of scale 规模经济

large scale 大规模的

on the scale of 按……的比例

scale up 按比例放大

hold / hoʊld /

v. 手持；容纳；举办；持有观点；保存（如在计算机中）

X-Rite eXact hand held 爱色丽 eXact 手持式（分光光度仪）

hold on 坚持；稍等

hold up 举起

hold off 推迟

equipment available 可用设备

Pantone ink 潘通油墨

Pantone book 潘通色卡

ink knives 墨刀

ink waste 浪费油墨

▶ [TRANSLATION]

第 14 单元　调墨

描述

调配恰当重量的潘通（专色）油墨，与参考标准进行匹配，安全（操作）和清洁将被评估。

工作	调配特殊的潘通专色
工作区域	工作台
可用设备	一系列潘通油墨 油墨秤 潘通色卡 计算器 墨刀 爱色丽 eXact 手持式（分光光度仪）

总工作时间	60 分钟，**不允许超时**
必要工作	选手需调配： 450 克 <u>PMS 302</u> 选手需调配： 450 克 <u>PMS 375</u> 你使用爱色丽 eXact 手持式（分光光度仪）测量每一种（专色）油墨的色彩。当你达到满意的 Lab 值时，呼叫裁判记录最终数值并提交一个你测量数值的（手工展墨）样张给裁判。 所有调配的（专色）油墨需要装罐。
工作完成后	完成（专色）油墨调配后，将（专色油墨）产品装入提供的容器中。 工作区域需保持清洁。
附加知识	不允许浪费油墨。 （操作）安全和清洁将被评估。

以上施工单选自第 44 届世界技能大赛印刷媒体技术项目的任务 4.1。

Part 1 Job Ticket

Unit 15 Ink Verification

Description

Verify mixed Pantone ink color to reference by using the printing press. Print specified quantity of test sheets. Print 2/0 adhesive stickers on a two color press using ink the competitors mixed in task nine. The competitors will be judged on print quality, color consistency and Lab accuracy. The competitors with the most color spots with the lowest Delta E will receive full marks. The competitors may adjust the ink formula but may not mix a new batch of ink.

Job	2/0 printing on two color press
Number of copies	200
Set-up waste	200
Total amount of paper	400
Colors	2/0 – PMS 302 and PMS 375
Paper size	Sheet size 35 cm×50 cm
Finished size before trimming	350 mm×500 mm
Finished size of each sticker	95 mm×95 mm and 57 mm×231 mm
Paper	Gloss Adhesive Paper
Ink	2/0 – PMS 302 and PMS 375 — the ink you mixed for the mixing completion
Machinery (Printing Press)	Heidelberg SX 52-2
Job status	You will punch and bend supplied offset plates — follow the instructions for the placement of plate and ink in correct printing unit.
Available time for the total task	1 hour, **NO overtime will be allowed**

During the printing

Competitors will achieve color, register, provide two signed "OK" sheets to the judges and stop the press. Give the technician your make ready sheet. Set counter to "0" and begin production with empty trolley.

The make ready sheet will be scanned by the X-Rite technician. Each color bar or spot will be measured.

Pull the last sheet and give it to the judges to be scanned.

After printing

Competitors will remove their job trolley from press and replace with an empty trolley. Competitors are to leave their plates on press.

The job ticket above comes from task 4.2 of the Print Media Technology project of the 44th WSC.

▶ [VOCABULARY & PHRASE]

consistency / kənˈsɪstənsi /	Delta E 色差
n. 一致性；黏稠度	two color press 双色印刷机
color consistency 色彩稳定性	finished size before trimming 裁切前尺寸
	adhesive paper 不干胶
formula / ˈfɔːrmjələ /	job trolley 收纸台板
n. 配方，处方；方案；公式	empty trolley 空的收纸台板
ink formula （专色）油墨配方	

▶ **[TRANSLATION]**

第 15 单元　油墨验证

描述

使用印刷机对调配好的潘通（专色）油墨色彩相对参考（色彩标准）进行验证。印刷规定数量的测试印张。选手使用在任务 9 调配的（专色）油墨，在双色印刷机上印刷 2/0 不干胶。选手（印张）的印刷质量、色彩稳定性和 Lab 值准确性将被评判。拥有最多色块及最小色差的选手得满分。选手可以调整（专色）油墨配方，但是不能重新调配（专色）油墨。

工作	在双色印刷机上印刷 2/0
印品数量	200
校版纸数量	200
纸张总量	400
色彩	2/0 – PMS 302 和 PMS 375
纸张规格	35 cm×50 cm
裁切前规格	350 mm×500 mm
不干胶成品规格	95 mm×95 mm 和 57 mm×231 mm
纸张	高光不干胶
油墨	2/0 – PMS 302 和 PMS 375，你为完成调墨（任务）调配的（专色）油墨
机器（印刷机）	海德堡 SX 52-2
工作状态	你对提供的胶印印版进行打孔和弯版。按照说明安装印版并注入（专色）油墨至正确的印刷机组。

工作总时间	1 小时，**不允许超时**
印刷中	选手进行色彩、套准调节，提交 2 张签字的 "OK" 样给裁判并停止印刷机。将你的校版纸交给技术人员。将计数器归 0，使用空的收纸台板开始生产。 校版纸将被爱色丽技术人员进行扫描。每一个色带色块都会被测量。 将（印品）最后一张抽出并交给裁判进行扫描。
印刷后	选手将收纸台板从印刷机上搬出并换上一个空的收纸台板。选手将印版留在印刷机内。

以上施工单选自第 44 届世界技能大赛印刷媒体技术项目的任务 4.2。

Part 1 Job Ticket

Unit 16 Cutting Exercises

Description

Prepare cutting plans and cutting printed sheets according to specifications.

Job	**Cutting Exercises**
Number of exercises	Exercise one will require trimming tent cards
Machinery (Paper Cutter)	Polar Paper Cutter
Signature size	297 mm × 420 mm
Finished size	204 mm × 291 mm
Available time for the total task	30 minutes to complete cutting for task 12 — **no overtime will be allowed**
Tasks to be done	1. Indicate cutting with a drawn line on one printed sheet. Number the lines to indicate the order of cutting. Checking of the trim size. 2. Carry out the cutting plan, program cutter and cut the tent cards according to the trim marks printed on the sheet. 3. Keep finished products in sequence order. 4. KEEP ALL WASTES, this is a live VDP job, any waste will have to be identified and reprinted.
After solving the tasks	Indicate to the judges when you have completed the exercise.
Additional knowledge	Perform cutting practices in a safe manner.

The job ticket above comes from task 4.4 of the Print Media Technology project of the 44th WSC.

▶ [VOCABULARY & PHRASE]

signature / ˈsɪgnətʃər /
n.（印刷）折标；签名；签署；鲜明特色，明显特征

signature size 签样规格

reprint / ˈriːprɪnt /
v. 重印；再版
n. 重印；翻版

reprint edition 重印版

paper cutter 切纸机

order of cutting 裁切顺序

sequence order 顺序

VDP 可变数据印刷，Variable Data Printing 的缩写

▶ [TRANSLATION]

第 16 单元　裁切练习

描述

根据规格（要求）制订裁切计划并裁切印张。

工作	裁切练习
练习数量	练习一需要裁切席卡
机器（切纸机）	波拉切纸机
签样尺寸	297 mm×420 mm
成品尺寸	204 mm×291 mm
工作总时间	30 分钟完成裁切任务 12，**不允许超时**

Part 1 Job Ticket

必要任务	1. 在一张印张上画线标识裁切（计划）。 将线进行编号，以标识裁切顺序。 核对裁切尺寸。 2. 实施裁切计划，对裁切机进行编程，根据印张上的裁切标记裁切席卡。 3. 对完成（裁切）的产品按序摆放。 4. 保留所有切下的废料，这是一个真实的可变数据工作，任何废料都会被鉴定或再印刷。
任务结束后	当你完成练习后示意裁判。
附加知识	以安全方式进行裁切练习。

以上施工单选自第 44 届世界技能大赛印刷媒体技术项目的任务 4.4。

Unit 17 Cutting Post Cards

Description

Like all print products, planning, efficiency and quality are of equal importance.

Marking

Measurement assessment will be used.

Time allotment

20 minutes

Task

The competitors will be given the post cards previously printed in the digital printing task. An efficient strategy for cutting the cards apart should be developed and followed when cutting the cards apart.

1. When the judge says to begin, take one printed sheet and draw lines where each cut is to be made.

(a) Finished size must be 100 mm×150 mm.

(b) You are not allowed to use block programming for this exercise. You must plan and program this exercise manually.

(c) Label the sequence of the cuts: 1, 2, 3, etc.

(d) Program Polar cutter to match your cutting plan.

2. Take the remaining stack of press sheets and use your cutting plan to cut them.

You must have 8 of each card.

3. When you complete, tell the judges and give them your cutting plan and finished cards.

The job ticket above comes from task 3.4 of the Print Media Technology project of the 45th WSC.

[VOCABULARY & PHRASE]

manually / ˈmænjuəli /
adv. 手动地；用手
manually actuated 人工启动的

block / blɑːk /
n. 大块；街区；大楼；障碍物；阻碍
v. 阻塞；遮住（视线）；封锁，挡住（去路）；阻挠；阻截；对……进行块操作
block programming 模块程序

[TRANSLATION]

第 17 单元　裁切明信片

描述

就像所有的印刷产品一样，计划、工作效率和（印品）质量同等重要。

评分

将使用测量评分进行评判。

时间分配

20 分钟

任务

选手将拿到之前在数字印刷任务中印刷的明信片。当裁切明信片各部分时，应该制定和遵守高效的裁切原则。

1. 在裁判宣布开始后，拿出一张印品进行画线，并标识每一个裁切位置。

(a) 成品尺寸为 100 mm×150 mm。

(b) 这个练习不可使用模块程序。这个练习你必须手动制订计划并编程。

(c) 标识裁切顺序：1、2、3 等。

(d) 根据你的裁切计划对波拉切纸机进行编程。

2. 按照裁切计划对剩余的印张进行裁切。

每种卡片需裁切 8 张。

3. 当你完成时，告知裁判并将你的裁切计划和成品卡片交给他们。

以上施工单选自第 45 届世界技能大赛印刷媒体技术项目的任务 3.4。

Part 1 Job Ticket

Unit 18 Simulation

Description

SHOTS simulation — errors come from not following standard press operating procedures for make ready. Competitors will be given one exercise with faults that are caused by a press operator not following standard operating procedures for make ready.

Job	**Print Simulation**
Number of copies	Exercise 1 requires no production copies be saved but the press sheets most be OK to save for task to be completed.
Machinery (Printing Press Simulation)	Sinapse Shots Sheetfed Offset Training Simulator
Job status	The simulator will be set as if it is a blank press.
Available time for the total task	1 hour to solve the exercise — **no overtime will be allowed**
During the printing simulation	Even though you are on a simulator, don't forget that the conditions are exactly the same as on the press.
	Check the paper, the ink and all the things you are used to and that you need to control before and during the job.
	Competitors will start with a blank press simulator with no ink setting. Competitors will have to set the ink fountains, achieve correct color and registration. There is only one situation with five defaults.

After solving an exercise

The system will ask you if you want to exit as soon as the exercise is completely finished "printing" good copies. That is very important. Notify the judges when you have completed.

If you attempt to restart an exercise during the exercise, you will receive zero for this exercise.

Additional knowledge

The competitors need to know how to use the simulator. The simulator technician will print a report and give it to the judges.

The job ticket above comes from task 4.8 of the Print Media Technology project of the 45th WSC.

▶ [VOCABULARY & PHRASE]

simulation / ˌsɪmjuˈleɪʃ(ə)n /	blank / blæŋk /
n. 模拟；仿造物；假装，冒充	adj. 空白的；毫无表情的；单调的；断然的
simulation game 模拟游戏	
simulation test 模拟试验	n. 空白处；（头脑或记忆）一片空白
	v. 成为空白
sheetfed / ʃiːt ˈfed /	
adj. 单张印刷的	
sheetfed offset 单张纸胶印（机）	blank press 未设置的印刷机

▶ **[TRANSLATION]**

第 18 单元　模拟

描述

SHOTS 模拟——故障来自准备（过程）没有遵守印刷机标准操作流程。选手需要完成一个故障练习，这些故障是印刷机操作员在准备（过程）中没有遵守印刷机标准操作流程导致的。

工作	印刷模拟
印品数量	练习 1 不需要进行印品印刷和保留，但是为完成任务，合格的印张是可以保留的。
机器（印刷机模拟器）	Sinapse Shots 单张纸胶印训练模拟器
工作状态	模拟器将被设置为一台未进行设置的印刷机。
工作总时间	练习时间 1 小时，**不允许超时**
模拟印刷时	尽管是在模拟印刷，不要忘记情况和在（真实）印刷机上是一致的。检查纸张、油墨以及所有你在印刷前或印刷中需控制和会使用的东西。 选手起始使用一台未经设置的模拟印刷机，且印刷机未进行油墨设置。选手需设置墨斗，进行正确的色彩和套准（调节）。一种模拟情况包含了 5 个故障。
练习结束后	当完成"印张印刷"练习后，系统会立即询问你是否要退出。这一步非常重要。当你完成时请示意裁判。 如果你在练习时试图重启练习，将被判 0 分。
附加知识	选手需知道如何使用模拟器。模拟器技术人员将打印一份（模拟器使用）报告并将其交给裁判。

以上施工单选自第 45 届世界技能大赛印刷媒体技术项目的任务 4.8。

Part 2

Dialogue

Unit 1 Who Is Who?

Jason: 选手

Lydia: 教练

Jason: Oh my, Lydia, I am going to the WSC next week but now I feel so **panic** about everything. You know, I have a slight social **phobia**.

Lydia: How come? Don't worry, everyone is very nice there. It's just a big **arena** for you to fight as a young **craftsman**.

Jason: I saw a lot of **acronyms** on the CR. I am afraid that I can't remember their names. I know I am a competitor, but I can't figure out what's the difference between SMT and SCM.

Lydia: SMT is short for Skill Management Team, while SCM stands for Skill Competition **Manager**. SMT **comprises** the Skill Competition Manager, Chief Expert, and Deputy Chief Expert.

Jason: So should I call them SCM in short or the full name Skill Competition Manager?

Lydia: You should read the full **title**, but when in written form, you can just write SMT. But when you meet them, you could just call sir or their first name if they allowed.

Jason: What would Chief Expert and Deputy Chief Expert do? What's the difference between Chief Expert and Expert?

Lydia: Experts would normally represent a member country or member region in the skill competition while the Chief Expert would be selected from the Experts and would manage the whole Experts' team; while Deputy Chief Expert would help Chief Expert to make decisions.

Jason: Or what if I am so nervous that I speak in my mother tongue?

Lydia: That's when you need your **interpreter**.

Jason: That sounds good. Do you know what is a WM?

Lydia: That's workshop manager, he is a person with **qualifications** and experience in their **accredited** skill who is responsible for workshop **installations** and security. If you need anything in the workshop, you can ask them for help.

Jason: Well, so who are CCDs?

Lydia: That's Competitions Committee **Delegate**, **basically**, you won't see them unless there is a technical problem with the test project. They would **normally oversee** the management of up to six skill competitions.

▶ [VOCABULARY & PHRASE]

panic / ˈpænɪk /
n. 恐慌，惊慌；大恐慌
adj. 恐慌的；没有理由的
v. （使）恐慌
panic attack 恐慌发作

phobia / ˈfoʊbɪə /
n. 恐怖，憎恶；恐惧（症）
silence phobia 怕安静

arena / əˈriːnə /
n. 竞技场
Allianz Arena 安联竞技场

craftsman / ˈkræftsmən /
n. 工匠；手艺人；技工
skilful craftsman 能工巧匠

acronym / ˈækrənɪm /
n. 首字母缩略词
acronym list 简写对照表

manager / ˈmænɪdʒər /
n. （公司、部门等的）经理；管理人员
project manager 项目经理

comprise / kəmˈpraɪz /
v. 包含；由……组成
be comprised of 由……组成

title / ˈtaɪt(ə)l /
n. 标题；（人名前表示地位、职业或者婚姻状况的）头衔；（电影、电视的）字幕
v. （给书籍、乐曲等）加标题；赋予头衔；把……称为
adj. 冠军的；标题的；头衔的
title bar 标题栏

interpreter / ɪnˈtɜːrprətər /
n. 口译者
simultaneous interpreter 同声传译员

qualification / ˌkwɑːlɪfɪˈkeɪʃn /
n. （通过考试或学习课程取得的）资格；

条件；限制；赋予资格
qualification certificate 资格证书
professional qualification 专业资格

accredit / əˈkredɪt /
v. 把……归于，归因于；委派；信任，正式认可；授权
accredited adj. 公认的；可信任的
accreditation n. 委派；信赖；鉴定合格

installation / ˌɪnstəˈleɪʃ(ə)n /
n. 安装；装置；就职
installation manual 安装手册

delegate / ˈdelɪɡeɪt /
n. 代表，会议代表；委员会成员
v. 授权，把……委托给他人；委派……为

代表，任命
technical delegate 技术代表
business delegate 业务代表
chief delegate 首席代表

basically / ˈbeɪsɪklɪ /
adv. 主要地，基本上
basically stable 基本稳定

normally / ˈnɔːrməlɪ /
adv. 正常地；通常地，一般地
normalization n. 标准化

oversee / ˌoʊvərˈsiː /
v. 监督；审查；俯瞰；偷看到，无意中看到
oversee the implementation of... 监督执行……

▶ [NOTES]

1. you know，语言学称之为"确认标记"，是母语为英语的人为交际用途而用的词。有时候，使用 you know 是想和交谈者有某种共识或者想知道对方是否同意他们的看法；有时候，他们用它来填补对话或讨论中的空白；而另一些时候，说 you know 是为了让发言者有时间思考下面该说些什么。

2. how come，意思是"为什么"，和 why 有一定的区别，how come 一般用于发生了什么事的场合。当有人对你说 how come 的时候，表示他不理解事情是怎么发生的，很希望得到你的解释。相比 why，how come 可以表达好奇和震惊的心情，语气更加强烈。此外，

how come 更口语化，多用于生活交际，后面接的是陈述句的语序。而 why 是书面和口语都可以使用的，使用范围更广，后面接的是疑问句。

3. CR，*Competition Rules*，《竞赛规则》；SMT，Skill Management Team，技能管理团队；SCM，Skill Competition Manager，技能竞赛经理；Chief Expert，首席专家；Deputy Chief Expert，副首席专家；WM，Workshop Manager，场地经理；CCD，Competitions Committee Delegate，竞赛委员会代表。

4. figure out，弄清楚，弄明白。

5. ...be short for...，是……的简称；是……的缩写。...stand for...，……代表……。此外，... for short 也表示"简称"。例如：The U.S. is short for the United States. /The United States is the U.S. for short.

6. comprise，组成。除了 comprise，compose, consist of, constitute 也有"组成"或"构成"的意思。comprise 在表示"构成"时，指的是"包括"或"覆盖"；compose 在表示"由……材料构成"时，常用于被动语态，在用于主动语态时，一般表示"构成"或"组成"；consist of 有"融合为一"的意思，而且主语是复数名词或集体名词；constitute 指所"构成"的事物在属性、特征或组织上与组成成分是一致的。

7. chief 和 deputy 分别指职位的正、副。除了 deputy，vice 作形容词时也有"副的"的意思。vice 一般用在较高的职务前作前缀来表示该职位的"副职"；deputy 常用于司局级、处级、科级前表示"副职"。

8. But when you meet them, you could just call sir or their first name if they allowed. 本句是一个将来时的虚拟句，从句的谓语动词用过去式，主句的谓语用 would/might/could + 动词原形。因此，if they allowed 作为从句，谓语动词用了过去式，而主句 you could just call sir or their first name 中用了 could + 动词原形。

9. mother tongue，母语。

10. be responsible for，为……而负责。

11. up to，至多。

▶ [TRANSLATION]

第1单元　人员介绍

杰森：　哦，天哪，莉迪亚，我下周要去世界技能大赛了，但是现在我对一切都感到非常恐慌。你知道的，我有轻微的社交恐惧症。

莉迪亚：怎么会？别担心，那里的每个人都很好。那只是一个让你作为年轻工匠去战斗的大舞台。

杰森：　我在《竞赛规则》上看到了很多缩写。我怕我记不住它们的名称。我知道我是一个选手，但我弄不清楚 SMT 和 SCM 有什么区别。

莉迪亚：SMT 是技能管理团队的缩写，SCM 则是技能竞赛经理的缩写。SMT 由技能竞赛经理、首席专家和副首席专家组成。

杰森：　那我应该称他们为 SCM 还是 Skill Competition Manager？

莉迪亚：你应该读全称，但如果是书面形式，你可以只写 SMT。但是当你遇到他们时，如果他们允许，你可以直接叫先生或他们的名字。

杰森：　首席专家和副首席专家会做什么？首席专家和专家有什么区别？

莉迪亚：专家通常代表一个参赛国家或地区参加技能比赛，而首席专家是从专家中选出并管理整个专家团队的；而副首席专家则帮助首席专家做出决策。

杰森：　我清楚了，但是如果我太紧张，说了我的母语怎么办？

莉迪亚：那你可以请翻译帮忙。

杰森：　听起来不错。你知道什么是 WM 吗？

莉迪亚：那是场地经理，一般是一个有资质、有经验的人，负责场地的设施和安全。如果你在赛区需要任何东西，可以找他们帮忙。

杰森：　好的，那 CCD 是谁？

莉迪亚：那是竞赛委员会代表，基本上，除非试题有技术问题，否则你不会看到他们。他们通常会监督最多六项技能比赛的技能问题。

Unit 2 WSC Culture

Jason: 选手

Lydia: 教练

Lydia: Wow, what's that on your **chest**?

Jason: That's a **pin** given by the **Brazilian** competitors.

Lydia: Oh, yes, that's part of the World Skill **traditional** culture.

Jason: Should I give the pins we have prepared to him?

Lydia: Yes, of course. The e**xchange** of pins is one of the most popular traditions in WSC. For each competition I would like to exchange as many pins as possible, and all the pins would be **pinned** on my **lanyards**. That's very popular during the competition days.

Jason: Speaking of which, could you please tell me what's the WSC culture and **heritage**?

Lydia: The **imagery** of WSC is youthful, energetic, and **inspiring**, and what they want to do is to inspire young people to **pursue** skills. They believe that skills can change lives, which **actually** would.

Jason: Of course, skilled people **deserve** that **recognition**. How many countries and regions are **involved** in WSC?

Lydia: As far as I know, now there are 85 members in the WSC, covering two thirds of the world's population.

Jason: Would they only compete with each other? Or are there any more **activities**?

Lydia: Of course it's far beyond competition. For example, there would also be a World Skills conference which would bring together leaders in education, government, business and industry from around the globe together to share best practice and learn about global **trends** and issues found in **Vocational** Education and Training (VET), skills demand, skills of the future as well as skills excellence and development.

Jason: OK, blablabla... that sounds **official** to me.

Lydia: Okay, I know you want to know something exciting. So you might hear about One School One Country and Skill Out.

Jason: No, I've never heard that before. I am all ears.

Lydia: One School One Country is a program. Schools in the host country are matched with a WorldSkills member country or region. In class, students learn about the values, traditions and culture of the WorldSkills member that they have adopted. Skill Out is just like a party where you can have some drink or grab some food when the competition is over.

Jason: Sounds nice, I will try it out.

Lydia: Yes, just go for it.

▶ [VOCABULARY & PHRASE]

chest / tʃest /
n. 胸，胸部；衣柜；箱子；〈英〉金库
on one's chest [口语] 心中有事，郁积在心，有心事

pin / pɪn /
n. 徽章；（尤指做衣服时固定布料用的）大头针，别针
v. （用钉子）钉住；压住；将……用针别住
rolling pin 擀面杖
pin something on someone [口语] 把某事的责任加在某人身上
on pins and needles 如坐针毡
pin down 使受约束；阻止

Brazilian / brəˈzɪliən /
adj. 巴西的；巴西人的
n. 巴西人
Brazilian market 巴西市场

traditional / trəˈdɪʃn(ə)l /
adj. 传统的；（活动）惯例的
traditional Chinese medicine 中医

exchange / ɪksˈtʃeɪndʒ /
n. 交换；交流
v. 交换；交易
exchange rate 汇率
exchange student 交流生
stock exchange 证券交易所
exchange for 交换
foreign exchange 外汇

lanyard / ˈlænjərd /
n. （用于悬挂哨子、证件等的）挂带
personal retention lanyard 个人保护系索

heritage / ˈherɪtɪdʒ /
n. 传统；遗产

cultural heritage 文化遗产
World Heritage Site 世界遗产保护区

imagery / ˈɪmɪdʒərɪ /
n.（艺术作品中的）像；意象；比喻；形象化
satellite imagery 卫星图
mental imagery 心理意象
visual imagery 视觉表象

inspire / ɪnˈspaɪər /
v. 激发（想法）；鼓舞；启示
inspire a generation 激励一代人
inspire sb. to do sth. 激励某人做某事

pursue / pərˈsuː /
v. 继续探讨；从事；追赶；纠缠
pursue one's studies 治学
the career I pursue 我理想的事业
pursue pleasure 寻欢作乐

actually / ˈæktʃʊəlɪ /
adv. 实际上；事实上
illuminations actually measured 实测照度

deserve / dɪˈzɜːrv /
v. 应受，应得
deserve to do sth. 值得干某事
deserve one's reputation 名不虚传

recognition / ˌrekəɡˈnɪʃ(ə)n /
n. 识别；承认
recognition system 识别系统
speech recognition 语音辨识

involve / ɪnˈvɑːlv /
v. 包含；牵涉；使陷于；潜心于
involve in 牵涉

activity / ækˈtɪvətɪ /
n. 活动；行动；活跃
economic activity 经济活动
catalytic activity 催化活性
physical activity 体育活动；体力活动
biological activity 生物活性

trend / trend /
n. 趋势，倾向；走向
v.〈美〉趋向，伸向
market trend 市场趋势
general trend 一般趋势
trend analysis 趋势分析

vocational / voʊˈkeɪʃən(ə)l /
adj. 职业的，行业的
vocational training 职业培训
vocational education 职业教育
vocational school 职业学校

official / əˈfɪʃ(ə)l /
adj. 官方的；正式的；公务的
n. 官员；公务员
official residence 官邸
official language 官方语言
official website 官方网站

host / hoʊst /
n. 主人；主持人
v. 主持；当主人招待

host country 东道国
host family 寄宿家庭
host city 主办城市

grab / græb /
v. 攫取；霸占；将……深深吸引
n. 攫取；夺取之物
up for grabs 大家有份
grab a bite 吃点东西
grab at 抓住，抓取

[NOTES]

1. pin，徽章。在世界技能大赛的观赛区，各国选手以交换各国世界技能大赛徽章象征友谊的传递，各个国家的徽章会凸显各个国家的特色。中国的徽章多以熊猫为显著特征，深受国外友人的欢迎。

2. speaking of which，一种口语惯用表达，意思为"正好说到这件事……"。

3. actually，口语中经常使用的高频词汇。有三种意思，①用来展示人们想法或观点的对立面，例如：No, I'm not a student. I'm a teacher, actually. 不，我不是一名学生。事实上，我是一名老师。②表示强调，例如：What did the expert actually say? 专家到底是怎么说的？③用于开始新话题，可以翻译为"其实"。Well actually, John, I called you for some advice. 嗯，约翰，其实我打电话给你是想寻求一些建议。

4. as far as I know / tell...，据我所知……。

5. far beyond...，远远超出……；远不止……。far 为程度副词，表示"远远……"。

6. Vocational Education and Training，职业教育与培训。需要区分 vocation（职业）和 vacation（假期）。

7. One School One Country，一校一队。"一校一队"项目于 2007 年日本静冈举办的第 39 届世界技能大赛首创，旨在鼓励当地青年踊跃参加比赛，且感受各国不同文化。自此以后，"一校一队"成为历届大赛的保留项目。"一校一队"项目是世界技能大赛的重要组成部分。

该活动旨在促进选手和举办国当地学生之间的交流，以期提高教学和职业培训的水平。

8. Skill Out，技能派对。它是比赛期间对所有专家、选手和翻译开放的一个小型餐会，便于他们交流与增进友谊。

9. I am all ears，洗耳恭听。相近的表达语句还有：be all eyes，目不转睛；be all heart，心地善良；be all mouth，只说不做；be all thumbs，手指不灵活。此外，还需要注意 tell me about it 的意思并非"快告诉我"或"告诉我关于它的事情"，而是"可不是嘛"，例如：His attitude is driving me crazy. Tell me about it. 他的态度真的快让我疯掉了。可不是嘛！

10. program，意思为"项目；程序"。这个词在英语中使用的频率和范围很广，有时候大学里的专业也可以用 program 来表示。

11. go for it，鼓励语。有"勇敢尝试，努力争取"的意思，还有"去吧"的意思，经常用于口语。

▶ [TRANSLATION]

第 2 单元　WSC 文化

莉迪亚：哇，你胸前的那个是什么？

杰森：　巴西选手给的徽章。

莉迪亚：哦，对，那是世界技能大赛传统文化的一部分。

杰森：　我应该把我们准备好的徽章给他吗？

莉迪亚：是的，当然。徽章交换是 WSC 最受欢迎的传统之一。每次比赛我都想换尽可能多的徽章，并且所有的徽章都会别在我的挂绳上。这是比赛期间非常受欢迎的活动。

杰森：　说到这里，你能告诉我 WSC 的文化和传统是什么吗？

莉迪亚：WSC 的形象是年轻、充满活力、鼓舞人心的，他们想要做的是激励年轻人追求技能。他们相信技能可以改变生活，而事实确实如此。

杰森：　当然，有技能的人应该得到这种认可。有多少国家和地区参加了 WSC？

莉迪亚：据我所知，现在 WSC 有 85 名成员，覆盖了世界三分之二的人口。

杰森：　只是涉及比赛吗？或者，还有什么活动吗？

莉迪亚：当然不止比赛了。例如，还有一个世界技能会议，将来自全球的教育、政府、商业和工业的领导者聚集在一起，分享各自的做法，了解职业教育和培训（VET）的全球趋势和问题、技能需求、未来技能、高精技能和技能发展。

杰森：　好吧……这听起来好官方。

莉迪亚：好的，我知道你想知道一些有意思的东西。那么，你可能听过"一校一队"和"Skill Out"。

杰森：　没有哎，我没听过。跟我讲讲。

莉迪亚："一校一队"是一个项目。东道国的学校与世界技能组织成员国或地区对接。在课堂上，学生了解世界技能组织成员国或者地区采用的价值观、传统和文化。"Skill Out"就像一场派对，比赛结束后你可以去喝点东西或吃点东西。

杰森：　听起来不错，我会试试的。

莉迪亚：是的，去试试吧。

Part 2 Dialogue

Unit 3 Skills Management Plan

Jason: 选手

Lydia: 教练

Jason: I've seen something named C-1, C1, C+1, what do they stand for?

Lydia: That's the **abbreviations** used to **indicate** days before, during and after the Competition. For example, "C1" is Competition day 1, "C-1" is one day before Competition day 1, and "C+1" is one day after the last day of the Competition.

Jason: I see. Are we **available** to make our own **arrangements besides** the Competition days?

Lydia: I wouldn't suggest that. The **preparation** work would normally starts from C-7 to C-1. The SMT would have a lot of meetings to make sure everything right.

Jason: So the competitors are free, right?

Lydia: No, you just have no access to the site. However, you need to **participate** some of the activities held by WSI, for example, C-3 would be the excursion day, C-2 would be the familiarization day and World Skills Champion Night while C-1 would be the One School One Country and Opening **ceremony**. So basically, you would be much occupied.

Jason: Wow, I didn't imagine that.

Lydia: Yes, that's just **routine**. The WSI has made well-organized **schedules** called SMP, Skills Management Plan. All you need to do is to **adapt** yourself.

▶ [VOCABULARY & PHRASE]

abbreviation / əˌbriːvɪˈeɪʃn /	indicate / ˈɪndɪkeɪt /
n. 缩写；缩写词	*v.* 表明；指出；预示；象征
English abbreviation 英语缩略语	indicate right 打右转向灯
abbreviation index 缩略词对照表	evidence indicate 迹象；预示
unit abbreviation 单位缩写	evidence-indicate 表明

available / əˈveɪləb(ə)l /
adj. 可获得的；可购得的
available material 可用材料

arrangement / əˈreɪndʒmənt /
n. 布置；整理；准备；安排
tour arrangement 旅行安排

beside / bɪˈsaɪd /
prep. 在……旁边；与……相比
adv. 在附近；况且，此外
beside the point 离题

preparation / ˌprepəˈreɪʃ(ə)n /
n. 预备；准备（指动作或过程）
surface preparation 表面处理
preparation for 建议

access / ˈækses /
n. 通道；进入；入口
v. 接近，使用；访问，存取（电脑文档）
adj. <美>（电视节目或时间等）对外公开的
access control 访问控制

participate / pɑːˈtɪsɪpeɪt /
v. 参与，参加；分享
public participate 公众参与
participate in 参加

excursion / ɪkˈskɜːrʒ(ə)n /
n. 短途旅行；远足
excursion train 旅行列车
excursion boat 游艇

familiarization / fəˌmɪliəraɪˈzeɪʃn /
n. 熟悉，精通；亲密
familiarization training 熟悉训练
familiarization visit 了解工作情况的访问
familiarization trip 熟悉之旅

ceremony / ˈserəmoʊnɪ /
n. 典礼，仪式
opening ceremony 开幕式
closing ceremony 闭幕式
award ceremony 颁奖典礼

routine / ruːˈtiːn /
n. 常规，惯例
adj. 常规的，例行的
v. 按惯例安排
routine business 日常事务
utility routine 实用程序
daily routine 日常生活
required routine 规定动作

schedule / ˈskedʒuːl /
n. 计划（表）；<美>（公共汽车、火车等的）时间表；一览表

v. 安排，预定；将……列入计划表或清单	adapt / ə'dæpt /
on schedule 按时	v. 使适应；改编
ahead of schedule 提前	adapt to 适合；改编；使适应
according to schedule 按照预定计划	adapt oneself to 使自己适应于
production schedule 生产计划	adapt for 使适合于；调剂

▶ [NOTES]

1. C-1, C1, C+1, 此处的字母 C 代表的是 competition，比赛日一般为 4 天，写作 C1、C2、C3、C4；比赛前的日期用 C- 来代替；比赛后的日期用 C+ 来代替。一般从 C-4 开始就已经有专家、翻译开始进行相关工作了，而选手也有自己相应的日程安排。

2. be available to，可被……利用或得到的。另一个相似的用法是 be available for，有效；有空做……；可供利用。be available for 后面加名词，be available to 后面既可加名词又可加动词。

3. besides 有两个词性，当介词时，是 "除……之外"；当副词时，是 "此外；以及" 的意思。注意，需要和 beside 进行区分，"beside" 是介词，有两个意思，"在……旁边" 和 "与……相比"。

4. have access to，使用；接近；可以利用；有权使用。give access to 接见；准许出入；向……开放；得以进入；可以接近。

5. excursion day，短途旅行。除 excursion 之外，journey，tour，travel，trip，voyage，expedition 与 cruise 这些名词均含有 "旅行" 之意。journey 是最普通的用词，侧重指时间较长、距离较远的单程陆上旅行，也指水上或空中的旅行。tour 含有 "最后返回出发地" 的意思，旅途中会停留游览点，距离可长可短，目的各异。travel 泛指旅行的行为而不特指某次具体的旅行，多指到远方作长期旅行，不强调直接目的地，单、复数均可用。trip 为普通用词，常指为公务或游玩作的较短暂的旅行，例如 business trip，出差。voyage 指在水上旅行，也可指空中旅行。excursion 为较正式用词，常指不超过一天的短时期娱乐性游玩，也可指乘火车或轮船往返特定景点的远足旅游。expedition 指有特定目的的远征或探险。cruise 主要指乘船的游览并在多处停靠。

6. familiarization day，熟悉日。比赛开始前，选手有 5～8 小时的时间准备其工位，同时对工具及材料进行检查和准备。在专家和场地经理的指导下，选手可利用该时间，熟悉设备、工具、材料与流程，并练习比赛中需要使用的设备，选手有权提问。若流程的难度太大，会有专门的指导人员就流程进行演示。

7. World Skills Champion Night，世界技能大赛冠军之夜。champion 为"冠军"之意，指在比赛中得第一名的人或动物、物品等，尤指体育项目比赛的冠军或锦标赛的优胜者。

8. basically，是频繁出现在英语母语者口语中的一个词，意思为"主要地，从根本上说，基本，总的说来"。但是，现在人们在口语中使用这个单词的时候，所说的事情跟 basic 基本无关，而是为了强调自己所说的是对的。

9. well-organized，由形容词/副词+过去分词构成的新词，此类型构成的词汇有很多，例如：newly-invented 新发明的；well-known 著名的；wide-spread 广泛流传的；highly-developed 高度发达的。

▶ [TRANSLATION]

第 3 单元　技能管理计划

杰森：　我看到一些名为 C-1、C1、C+1 的东西，它们代表什么？

莉迪亚：那是比赛前、比赛中和比赛后几天的缩写。例如，"C1"是指比赛第 1 天，"C-1"是指比赛前一天，"C+1"是指比赛后一天。

杰森：　我明白了。除了比赛日，我们是否可以自行安排？

莉迪亚：我不建议这样做。准备工作通常从 C-7 开始到 C-1 结束。SMT 团队会召开很多会议以确保一切进行顺畅。

杰森：　所以选手没事做，对吗？

Lydia：　不，你们只是没办法进入场地。但是，你需要参加 WSI 举办的一些活动，例如，C-3 是游览日，C-2 是熟悉日和世界技能大赛冠军之夜，而 C-1 是"一校一队"活动和开幕式。所以基本上你都会很忙。

杰森：　哇，我没想到。

莉迪亚：是的，这都是一些例行活动。WSI 制订了安排妥贴的计划，称为 SMP，即技能管理计划。你需要做的就是让自己适应。

Unit 4 Meeting New Friends

Tiago: 巴西选手

Jason: 中国选手

Alexander: 哈萨克斯坦选手

Tiago: Hi, you are Jason, right?

Jason: Yes! And you are?

Tiago: I am Tiago, the Brazilian competitor.

Jason: Oh yeah! I've seen your name on the **roster**. Nice to see you...eh... How should I **pronounce** your name?

Tiago: Ti-a-go, Tiago.

Jason: OK, got it! Tiago. Now I know a **Portuguese** word.

Tiago: Good for you! Haha. How long have you been here?

Jason: Two days. I've just got everything settled down.

Alexander: Hi, there.

Tiago: Wow, hi, Alexander! Where did you come from? You **creep** me out! Hey, Jason, this is my friend Alexander, the Kazak competitor. This man speaks the perfect English.

Jason: Oh, hi, Alexander. I'm Jason.

Alexander: Good to see you, Jason. I saw you two talking so I guess that I can make myself known here.

Tiago: Yes, you did it. Are we going to **assemble**? I see the other competitors coming.

Jason: Could you please introduce them to me? I am just new here.

Alexander: Of course, you can count on him. Tiago is quite a star here.

Tiago: Don't make fun of me. No worries, Jason. I'll introduce everyone to you. They are the competitors from Austria, Belgium, Estonia and Switzerland.

Jason: You are a great help.

Tiago: Come with me. Let's say hi to them.

▶ [VOCABULARY & PHRASE]

roster / ˈrɑːstər /

n. 花名册；执勤人员表

duty roster 轮值表

shift roster 轮班；轮更表

pronounce / prəˈnaʊns /

v. 发音；宣判；断言；发表意见，表态

pronounce on 对……发表意见

Portuguese / ˌpɔːrtʃʊˈgiːz /

adj. 葡萄牙的；葡萄牙人的

n. 葡萄牙语；葡萄牙人

Portugal n. 葡萄牙

settle / ˈset(ə)l /

v. 解决（分歧，纠纷等）；定居；沉淀；（地面或建筑）下陷

n. 有背长椅

settle down 定居

settle down to 专心致力于

settle up 付清

creep / kriːp /

v. 爬行；蔓延；慢慢地移动；（由于恐惧、厌恶等）起鸡皮疙瘩

n. 爬行；〈非正式〉毛骨悚然的感觉

creep in 悄悄混进

assemble / əˈsemb(ə)l /

v. （使）集合，（使）聚集；装配；收集

assemble line 装配线

▶ [NOTES]

1. got it, 表示"弄清楚了，弄明白了"。类似的表达方式还有 OK, all right, sure。好的，知道了，没问题。Fair enough. 说的对。That makes sense. 有道理。I know what you mean. 我明白你的意思。

2. good for you, 表示"为某人感到开心，感到高兴"，有些情况下可以与 congratulations, good job, well done, brilliant, bravo, you deserve it, proud of you 等短语互换使用。在某些场景下，这句口语也表示揶揄，类似中文中的"呵呵"。

3. get everything settled, 一切安顿下来。

4. where did you come from, 意为"你刚去哪里了"，注意和 where do you come from

的区别，后者意思是"你是哪里人"。

5. creep sb. out，吓着某人。creep out 的意思是"爬出去"。起了鸡皮疙瘩的表达为：I've got some goosebumps.。

6. speak the perfect English，说一口流利的英语。

7. I am just new here. 我是一个新人。

8. count on sb. to do sth.，指望某人做某事。

9. Don't make fun of me. 不要开我玩笑。make fun of sb.，和某人开玩笑。

10. no worries，不客气；不用担心；无忧无虑。这是英国人、澳洲人在回复他人感谢时的口头语。除了表示"不客气"，在回复对方的道歉时也指"别在意，没关系"。注意，此处 worries 不能用单数。回复 Thank you 的常用表达还有：Don't mention it./It's my pleasure./No problem./ Any time. 等等。

11. You are a great help. 你帮了大忙。

▶ [TRANSLATION]

第 4 单元　　遇见新朋友

蒂亚戈：　嗨，你是杰森，对吧？

杰森：　　是的！你是？

蒂亚戈：　我是蒂亚戈，巴西选手。

杰森：　　哦耶！我在名册上看到了你的名字。很高兴见到你……呃……你的名字要怎么念呀？

蒂亚戈：　蒂——亚——戈，蒂亚戈。

杰森：　　好的，知道了！蒂亚戈。现在我学会了一个葡萄牙语单词。

蒂亚戈：　很不错！哈哈。你来这里多久了？

杰森：　　两天。我刚刚安顿下来。

亚历山大：你们好呀。

蒂亚戈：　哇，你好，亚历山大！你刚去哪里了？你吓到我了！嘿，杰森，这是我的朋友亚历山大，哈萨克斯坦选手。这个人说的英语是一流的。

杰森： 哦，亚历山大，你好。我是杰森。

亚历山大： 很高兴见到你，杰森。我看到你们两个在说话，所以我也想过来让你认识一下我。

蒂亚戈： 好的，认识了。我们要集合了吗？我看到其他选手过来了。

杰森： 你能介绍我认识一下他们吗？我是个新人。

亚历山大： 当然，全得靠他。蒂亚戈在这很有名的。

蒂亚戈： 别取笑我。不用担心，杰森。我向你介绍一下他们。他们是来自奥地利、比利时、爱沙尼亚和瑞士的选手。

杰森： 你帮大忙了。

蒂亚戈： 跟我来。让我们跟他们打个招呼。

Unit 5 If You've Got Any Question?

Ben: 专家

Jason: 选手

Sam: 首席专家

*When facing a problem, Jason **raises** his hand up...*

Ben: What can I do to help?

Jason: I don't get this sentence, it doesn't make sense in Chinese. Is it possible to have my interpreter/expert here?

Ben: Sure, no problem.

Several minutes later, Jason puts up his hand again.

Ben: What's the matter, young man?

Jason: I met some problems with this **folding** machine/press/feeder/cutting machine, can I stop the counter?

Ben: Sorry, you can't.

Jason: But there should be something wrong with it. If it's not stopped, it wouldn't be **fair** for me.

Ben: I can't decide it. I'll bring the Chief Expert here to see what he can do for you.

Jason: Thank you, sir.

Sam: What's wrong, dear?

Jason: I believe there's something wrong with the machine. I am asking for a time **suspension**.

Sam: Let me see... Eh, maybe we can give you an extension of time.

Jason: And also sir, would it be ok if I perform in this way?

Sam: It's up to you. / It's your own call.

Jason: But what if I get a **deduction** of marks?

Sam: Everything is possible. It's still your own call. I can't tell you now. But maybe later we can have a discussion on it.

▶ [VOCABULARY & PHRASE]

raise / reɪz /

v. 提高（数量、水平等）；筹集；养育；升起；饲养；种植

n. 高地；上升；＜美＞加薪

raise oneself 长高

raise money 集资；募捐

get a raise 得到加薪

raise capital 筹集资本

fold / foʊld /

v. 折叠；可折叠

n. 褶层；褶痕

fold increase 成倍增加

fold line 折线；折纹；返折线

fair / fer /

adj. 公平的；美丽的；（肤色）白皙的

adv. 公平地；直接地；清楚地

v. 转晴

n. 展览会

vanity fair 名利场

fair use 合理使用

fair play 公平竞赛

suspension / sə'spenʃ(ə)n /

n. 悬浮；暂停；停职

in suspension 悬浮中

deduction / dɪ'dʌkʃn /

n. 扣除，减除；推论；减除额

salary deduction 罚薪

tax deduction 减免税款

▶ [NOTES]

1. I don't get this sentence. 我不明白这句话。get sth. 多用于口语中，表示"明白某事，懂了……"。

2. make sense，弄清楚，弄明白；有道理，可以讲得通；易于理解，表达清楚。

3. Is it possible to have my interpreter / expert here 也可以用 Sorry, I need my interpreter here 来表达。Is it possible to... 是委婉地表示请求的方法。除此之外，可以用 Is there any way of doing sth./ Is it ok to do sth. / Is there any chance that you could... / Is it convenient for you to... 以及下文中出现的 would it be ok to do sth. 来委婉地表达请求。

4. There should be something wrong with it. 可能哪里出了问题。这也是委婉提出问题的

一种表达方式。

5. to see what he can do for you，看看他能为你做些什么。

6. give you an extension/ask for an extension，给你延长时间 / 请求延长时间。

7. It's up to you. / It's your own call. 都是"你来决定"的意思。除此之外，可以用 You are the boss. 来表示"你说了算"。

▶ [TRANSLATION]

第 5 单元　如果你有任何问题？

当遇到问题时，杰森举起了手……

本：　　我能帮你什么吗？

杰森：　我不明白这句话，在中文里这说不通。我的翻译 / 专家可以来一下吗？

本：　　当然，没问题。

几分钟后，杰森再次举起了手。

本：　　怎么了，年轻人？

杰森：　这台折页机 / 印刷机 / 飞达 / 裁纸机出现了一些问题，我可以停止计时吗？

本：　　对不起，不能。

杰森：　但它应该是有问题的。如果你不停止计时，这对我来说是不公平的。

本：　　我自己决定不了。我带首席专家过来，看看他能为你做些什么。

杰森：　谢谢你，先生。

山姆：　亲爱的，怎么了？

杰森：　我觉得这台机器有问题。我要求暂停一下时间。

山姆：　让我看看……呃，也许我们可以给你延长时间。

杰森：　还有先生，我这样操作是可以的吗？

山姆：　这取决于你。/ 这是你自己的决定。

杰森：　但如果我被扣分怎么办？

山姆：　这都是有可能的。这仍然要你自己决定。我现在不能告诉你。但也许等会儿我们可以讨论一下。

Unit 6 Dispute Resolution

Sam: 专家

Jason: 中国选手

David: 技能竞赛经理

Alexander: 哈萨克斯坦选手

Sam: Hey!!! Young man, what are you doing?

Jason: I am giving a gift to my friend from Costa Rica. Is there a problem, sir?

Sam: Please open it up. We need to check what's in it.

Jason: It's just a pen drive, with 1 **terabyte**. I hope that when he uses this pen drive, he can think of me.

Sam: Sorry, young man. Competitors, experts and interpreters give a USB pen drive or other portable **storage** devices in any form as "give-aways" etc. to or from other **participants** must store this in their own locker immediately. I've sent you an e-mail before you came here. It's **strictly forbidden** to give a USB pen drive in the workshop.

Jason: I am sorry sir. I didn't notice it.

Sam: Young man, you should know that if a competitor breaches the skill-specific rules or Competition Rules, he or she will receive either 600 points or 5 points less than the lowest score of the lowest scoring competitors across all skills competitions, whichever is lowest.

Jason: I am sorry sir. It's not my **intention** to do that. I'll bring this pen drive out of the workshop **immediately**.

Sam: Ok, never do that again.

During the competition...

David: What's the problem?

Sam: The competitors continued to work after the count down.

Alexander: I can fix this, please give me some **extra** time.

David: Sorry young man. You're against the *Competition Rules*. We might **deduct** your scores.

If you've got any question, you can contact your Technical Delegate for **disputing** the resolution.

▶ [VOCABULARY & PHRASE]

terabyte / ˈterəbaɪt /
n.（计算机）万亿字节，兆兆字节（信度量度单位）
one terabyte 1TB

storage / ˈstɔːrɪdʒ /
n.（信息的）存储；仓库；贮藏所
data storage 数据存储
storage system 存储系统
energy storage 蓄能
storage tank 储油罐

participant / pɑːrˈtɪsɪpənt /
n. 参与者，参加者
adj. 参与的
participant observation 参与观察

strictly / ˈstrɪktlɪ /
adv. 严格地；完全地；确实地
strictly confidential 绝对机密

forbidden / fərˈbɪdn /
v. 禁止；妨碍；禁止……进入；阻止（forbid 的过去分词）
adj. 不被允许的，被禁止的；禁戒的

The Forbidden City 紫禁城；故宫

breach / briːtʃ /
n. 违背，违反；缺口
v. 违反，破坏；打破
breach of contract 违约
data breach 数据泄露

intention / ɪnˈtenʃ(ə)n /
n. 意图；目的
purchase intention 购买意愿
original intention 初衷

immediately / ɪˈmiːdɪətlɪ /
adv. 立即，立刻；直接地
conj. <英> 一……就
immediately following 紧跟着
reply immediately 秒回

extra / ˈekstrə /
adj. 额外的；另外收费的；特大的
n. 额外的事物；另外收费的事物；（非击球所得的）附加分
adv. 额外；特别地，格外地
pron. 额外的东西（尤其指钱财）

extra time 额外时间；加赛时间	deduct from 从……中减去
extra money 额外的钱	deduct wages 扣工资
extra work 加班；额外工作	
extra large 加大码的	dispute / dɪˈspju:t /
extra charge 附加费	n. 辩论；争吵；意见不同
	v. 辩论；对……进行质疑；争夺；抵抗（进攻）
deduct / dɪˈdʌkt /	dispute resolution 调解纠纷
v. 扣除，减去	in dispute 在争论中
deduct money 扣钱；扣款	labor dispute 劳动纠纷

[NOTES]

1. I am giving a gift，赠送礼物。赠送礼物是世界技能大赛的一个传统。参赛人员准备各自国家的特色小礼物互相赠送，利于增强友谊。

2. Costa Rica，哥斯达黎加。

3. I didn't notice it. 我没注意到。口语中经常用到的"我没看到"也可以用这句话表示。

4. breach 通常被翻译为"违约"或"违反"，在法律中较为常用。常用的短语有：breach of contract，违约；breach of confidence，泄露秘密；breach of faith，背信；breach of the condition，违反合同条件（条款）。注意 breach 和 break 的区别，break 就是一般的"损坏，破坏"，例如：break the glasses，眼镜碎了。

5. skill-specific rules，技能特定规则。在竞赛期间，项目经理们必须制定特定的适用于竞赛项目的规则。

6. It's not my intention. 这不是我的初衷。

7. pen drive，优盘，此种表述方式出现在 2019 年的技能特定规则中。而口语中表示优盘的英文说法还有：USB flash disk, U disk, flash disk；而移动硬盘为 mobile disk。

8. fix 的用法很广，除了表示"修理、解决"，还有"确定""安排""准备""决定"的意思。例如：fix supper，准备晚饭；fix the price，确定价格；fix sb. with a job，给人安排工作；fix a date for the meeting，确定开会时间。

9. Technical Delegate,技术代表。每个成员国或者成员地区会有一名技术代表，一般遇到重大问题时，该国（地区）专家会提请问题到技术代表处，由技术代表出面进行解决。

10. dispute resolution，调解纠纷；解决争议。一般出现争议时会按照《竞赛规则》的处理方式和流程解决。

▶ [TRANSLATION]

第6单元　调解纠纷

山姆：　　喂！！！年轻人，你在做什么？

杰森：　　我要给来自哥斯达黎加的朋友送礼物。有问题吗，先生？

山姆：　　请打开它。我们需要检查里面有什么。

杰森：　　它只是一个容量为1TB的优盘。我希望他用这个优盘的时候能想起我。

山姆：　　对不起，年轻人。选手、专家和翻译向其他参赛者赠送或从其他参赛者处获得优盘或其他任何形式的便携式存储设备作为"赠品"时，必须立即将其存放在自己的储物柜中。在你来之前，我已经给你发了一封电子邮件。严禁在赛场赠送优盘。

杰森：　　对不起，先生。我没有注意到。

山姆：　　年轻人，你应该知道，如果选手违反了特别技能规则或比赛规则，他要么只能拿到600分，要么是比所有技能比赛中得分最低的选手的最低分再少5分的分数，会是全场最低分。

杰森：　　对不起，先生。这不是我的本意。我马上把这个优盘带出赛场。

山姆：　　好吧，不要再那样做了。

比赛期间……

大卫：　　有什么问题？

山姆：　　倒计时结束后，选手还在继续工作。

亚历山大：我可以解决这个问题，请再给我一点时间。

大卫：　　对不起，年轻人。你已经违反了《比赛规则》。我们可能会扣除你的分数。如果你有任何问题，可以联系你的技术代表来解决争议。

Unit 7 How to Get to the Competition Venue?

Jason: 中国选手

Lydia: 教练

Jason: I have just received an **accommodation scheme** which introduces the World Skills Village.

Lydia: Wow, is it good?

Jason: Seems quite well. We've got **accreditation** center, catering services, meeting rooms, **physical security perimeter**, **shuttle** bus, **electro-car** transportation, swimming pool, fitness center and even self-service laundry.

Lydia: Have you checked the navigation map? Make sure you can find your workshop and the shuttle bus parking lot.

Jason: How can I **locate** the right skill workshop?

Lydia: You need to find the skill cluster first, and then the specific skill venue would be in each cluster. OH! You should also make sure you can find the rest room. You can take advantage of familiarization day to do that.

Jason: Thank you for your **consideration**!

Lydia: Oh! You would also get you accreditation badge, which would show your **identity**, so you can move around with this card for shuttle bus and meal.

Jason: Is there anything else I should pay attention to?

Lydia: Yes, keep pace with your expert and interpreter in case of some emergencies. Normally, the competitors would only be in the Expo or in the World Skills Village.

Jason: Ok. Got it. How should I get to the **airport** or the **railway** station?

Lydia: The shuttle bus would be arranged for you to the airport and railway station.

▶ [VOCABULARY & PHRASE]

accommodation / əˌkɑːməˈdeɪʃ(ə)n /
n. 住处，膳宿；调节；和解
accommodation cost 住宿费用

scheme / skiːm /
n.（政府或其他组织的）计划；组合；体制；诡计
v. 搞阴谋；拟订计划；策划
pension scheme 退休金计划
sampling scheme 抽样方案
color scheme 配色方案

accreditation / əˌkredɪˈteɪʃn /
n. 委派；信赖；鉴定合格
accreditation body 认可机构
accreditation system 认可体系

physical / ˈfɪzɪk(ə)l /
adj. 物理的；身体的
n. 体格检查
physical object 物体

security / sɪˈkjʊrəti /
n. 安全，安全性；保证；证券；抵押品
network security 网络安全
Security Bureau 安全局
homeland security 国土安全

Security Council 安理会
security guard 保安员

perimeter / pəˈrɪmɪtər /
n. 周长；周界
security perimeter 安全边界

shuttle / ˈʃʌt(ə)l /
n. 航天飞机（space shuttle 的简称）；穿梭班机、公共汽车等
v. 穿梭，往返
space shuttle 航天飞机
shuttle bus 班车

electro-car / ɪˈlektroʊ- kɑːr /
n. 电动汽车

laundry / ˈlɔːndri /
n. 洗衣店；洗衣房
laundry detergent 洗衣粉
laundry service 洗衣服务
laundry bag 洗衣袋
laundry room 洗衣房

navigation / ˌnævɪˈgeɪʃ(ə)n /
n. 航行；航海
satellite navigation 卫星导航

navigation bar 导航栏

navigation equipment 导航设备

locate / ˈloʊkeɪt /
v. 确定……的位置；探明
locate the fault 找出问题

cluster / ˈklʌstər /
n. 群；簇；丛；串
v. 群聚；丛生
a cluster of 一群
cluster sampling 分组取样
star cluster 星团

consideration / kənˌsɪdəˈreɪʃ(ə)n /
n. 考虑；原因；关心；报酬
under consideration 在考虑中

badge / bædʒ /
n. 徽章；证章；标记
v. 授给……徽章
arm badge 袖章
name badge 姓名牌
detective boys badge 侦探徽章

identity / aɪˈdentətɪ /
n. 身份，本体；个性，特性；同一性，一致；恒等运算，恒等式
an identity of interests 利益共同体

pace / peɪs /
n. 一步；步速；步伐；（移动的）速度
v. （因担忧、紧张或不耐烦而）踱步，踱步于；缓慢而行
a change of pace 节奏变换
keep pace with 保持同步
to set the pace 控制进攻节奏
pace maker 领跑

airport / ˈerpɔːrt /
n. 机场；航空站（港）
airport terminal 航站楼

railway / ˈreɪlweɪ /
n. （英）铁路；（其他交通工具使用的）轨道；铁道部门
v. 乘火车旅行
railway station 火车站

▶ [NOTES]

1. accommodation scheme，住宿方案。

2. quite，"非常；相当，很"，可以和 very 互换。quite 与一些表示"完整""完全"的意思的词或词组，如 all right, certain, determined, empty, finished, ready, right, sure, wrong 等连用，或与一些具有强烈感情色彩的形容词、副词，如 amazing, extraordinary, horrible, perfect 等连用时，它的意思是"完全地""全然地"。

3. accreditation center，认证中心。认证中心一般是办理通行证或者住宿的地方。catering services，餐饮服务。physical security perimeter，安检系统，也可以用 safe examination system 来表示。shuttle bus，班车，一般用来运送工作人员往返赛场、选手村或者机场。fitness center，健身中心。self-service laundry，自助洗衣房。

4. navigation map，导航地图。

5. parking lot，停车场。

6. skill cluster，技能大类的场地。世界技能大赛的比赛项目有 6 大类，分别为结构与建筑技术、创意艺术和时尚、信息与通信技术、制造与工程技术、社会与个人服务、运输与物流。

7. specific skill venue，特定技能场地。venue 多指"（体育）场馆；会场；场地；集合地点"。place 指的是较为广义的"位置"。而 location 多指方位位置。

8. rest room，厕所、洗手间。不同国家和地区对厕所的叫法也不同。比较常见的是 toilet, wash room 和 rest room。注意：bathroom 一般指的是带淋浴、洗澡功能的洗漱间；中国常用的 WC 其实是 water closet 的首字母缩写，在国外一般不使用。对于休息室，不同功能，叫法也不相同，例如，机场、酒店的休息室称为 lounge，公司员工的休息间叫 staff room，酒店的休息大堂为 lobby。

9. Thank you for your consideration. 或者 Thank you for your concern.，感谢您的考虑。

10. accreditation badge，通行证。

11. keep pace with，与……齐步前进，与……步调一致，跟上。

12. in case of，万一，以防。in case of 中的 of 是介词，介词后面只能接带名词性质的词，比如名词、代词等。in case 的意思是"万一"，它是连词，引导条件状语从句，后面是一个完整的句子。

[TRANSLATION]

第 7 单元　如何去赛场？

杰森：　我刚刚收到一份介绍世界技能选手村的住宿方案。

莉迪亚：哇，怎么样？

杰森：　看起来还不错。我们住的地方有认证中心、餐饮服务、会议室、安检系统、班车、电动汽车接送、游泳池和健身中心，甚至自助洗衣房。

莉迪亚：你看过导航地图吗？你要确保可以找到赛场和班车停车场。

杰森：　我怎么去赛场？

莉迪亚：你需要先找到技能大类场馆，然后在每个场馆中找到具体的技能场馆。哦！你还应该确保自己能找到洗手间。你可以利用熟悉日的机会来做到这一点。

杰森：　谢谢你给我这么贴心的考虑！

莉迪亚：哦！你还会拿到通行卡，上面有你的身份信息，因此你可以带着通行卡四处走动，乘坐班车和吃饭。

杰森：　还有什么我需要注意的吗？

莉迪亚：是的，在某些紧急情况下，要与专家和翻译保持同步。通常，参赛者的活动地点都在场馆或选手村。

杰森：　好的，明白了。我应该怎么去机场或火车站？

莉迪亚：会有安排好的班车载你到机场和火车站。

Part 2 Dialogue

Unit 8 Publication of Results

Lydia: 教练

Jason: 选手

Lydia: The competition days are over. How is everything going?

Jason: I am more confident with my measurement marks. I am not so sure with the judgment marks.

Lydia: Yes, especially when the WSC is increasing the judgment marking.

Jason: Do you know how could they make sure that it's a fair play?

Lydia: The mark **awarded** is **calculated** from the scores awarded by three experts in the marking team. The score must be between 0 and 3. All three experts **display** their flash cards at the same time when directed by the experts coordinating the recording of scores. Where the **range** of scores awarded for an **aspect** is greater than 1, experts must re-score that aspect.

Jason: So, what's the **benchmark**?

Lydia: The benchmarks would be provided in the *Marking Scheme* (and recorded on the marking forms) where it would appear as a context for these standards and act as a reference point for the marking team.

Jason: Will I get the final result on a paper or a poster?

Lydia: A handwritten mark sheet must be created to record the finally agreed scores. This is used for data entry into the CIS and kept to provide an audit trail. The assessment and CIS entry must **occur** before 22:00 on C4 and approval and sign-off must be received by the CIS team by 10:00 on C+1. When CIS mark entry has been locked, a PDF of all the marking forms for the specified marking day – including the Mark Summary Form – would be created. And then all the experts would review the PDF of the results for their **compatriot** competitors against the handwritten marks.

Jason: And then we would get all the marks results?

Lydia: No. The official results will be posted to the WorldSkills International website as medal winners are **announced** at the Closing Ceremony (C+1), **simultaneously**.

113

Jason: So they just want to give us a surprise?!

Lydia: Yes. You wouldn't know the results <u>on the last minute</u>.

Jason: Oh! I dare say that I would **faint** when they publish the results.

▶ [VOCABULARY & PHRASE]

award / əˈwɔːrd /
n. 奖；（收入的）增加；（赔偿）裁定额；（毕业证书等的）授予；奖学金
v. 把（某物）授予（某人）；把（合同、佣金）给（人、组织）
Hugo Award 雨果奖
literary award 文学奖

calculate / ˈkælkjuleɪt /
v. 计算；以为；预测，推测
calculate area 计算面积

display / dɪˈspleɪ /
n.（计算机屏幕上的）显示；炫耀
v.（计算机）显示；表现；陈列
adj. 展览的；陈列用的
LCD display 液晶显示器
on display 展出

range / reɪndʒ /
n.（变动或浮动的）范围；幅度；排；山脉
v.（在一定范围内）变动；漫游；射程

达到；使并列
a range of 一系列
full range 全距
in the range of 在……范围之内

aspect / ˈæspekt /
n. 方面；方向；形势；外貌
aspect ratio 纵横比
aspect angle 视界角
environmental aspect 环境因素

benchmark / ˈbentʃmɑːrk /
n. 基准；标准检查程序
benchmark interest rate 基准利率
benchmark system 基准系统
benchmark test 基准测试

occur / əˈkɜːr /
v.（尤指意外地）发生；出现；存在
occur as 以……的形式出现
occur for 发生在……时候

compatriot / kəmˈpeɪtrɪət /

n. 同胞；同国人	simultaneously / ˌsaɪm(ə)lˈtemɪəslɪ /
adj. 同胞的；同国的	adv. 同时地
overseas compatriot 海外同胞	develop simultaneously 同时发展
announce / əˈnaʊns /	faint / feɪnt /
v. 宣布；述说；预示；播报；通报……的到来；宣布竞选公职	adj. 模糊的；头晕的
	v. 昏倒；变得微弱
announce results 公布成绩	n. 昏厥，昏倒
announce the opening of 宣布……的召开	faint scent 清香
announce to sb. 通知某人	faint with 由于……原因而衰弱

▶ [NOTES]

1. How is everything going? 事情进展得如何？在书信体格式中，指的是"最近怎么样"。

2. be confident with sth., 对……感到自信。

3. measurement marks，测量评分；judgment marks，评价评分。这是世界技能大赛所有比赛项目分数的两大类。测量评分为客观分，评价评分则为主观分。

4. flash cards，卡片，也指闪存卡、教学用的单词卡等。

5. Where the range of scores awarded for an aspect is greater than 1, experts must re-score that aspect. 此处的 where 引导条件状语从句，意思为"在……的情况下"。

6. *Marking Scheme*，《评分方案》。

7. CIS，Competition Information System，竞赛信息系统。世界技能大赛最终采取的分数制度为世界技能分制，目的是让各竞赛项目之间可以进行比较。成绩按 100 分制打分，再按照世界技能的分制由 CIS 系统进行标准化。此程序将把每个竞赛项目中的 700 分设定为中位数。

8. audit trail，审计跟踪。指用来进行审查的详细记录。

9. Mark Summary Form，评分总表。

10. on the last minute，在最后一刻，在最后一分钟。

▶ [TRANSLATION]

第8单元　结果发布

莉迪亚：比赛日结束了。你还好吗？

杰森：我对自己的测量评分更有信心。我对评价评分不太确定。

莉迪亚：是的，尤其是当 WSC 正在增加评价评分的比重时。

杰森：你知道他们如何确保这是一场公平的比赛吗？

莉迪亚：评分是由评分小组中的三位专家的评分计算得出的。分数必须在0到3之间。在协调记录分数的专家的指导下，所有三位专家同时展示他们的评分卡。如果某个项目的得分差值大于1，则专家必须对该项目重新评分。

杰森：那么，基准是什么？

莉迪亚：基准将在《评分方案》中提供（并记录在评分表中），它将作为这些标准的背景出现，并作为评分团队的参考点。

杰森：我会在纸上或海报上看到最终结果吗？

莉迪亚：专家们必须手写评分表来记录最终商定的分数。这是用来将数据输入 CIS 并保留下来以提供审计跟踪。评分和 CIS 成绩录入必须在 C4 的 22:00 之前完成，并且必须在 C+1 的 10:00 之前收到 CIS 团队的批准和签字。当 CIS 分数录入被锁定时，将生成特定评分日的所有分数表格的 PDF 版本，包括评分总表。然后，所有专家将对照手写评分来校对他们本国选手分数的 PDF 版本。

杰森：然后我们会得到所有的分数结果？

莉迪亚：不会。闭幕式（C+1）宣布奖牌获得者的同时，官方结果将会被发布到世界技能组织的网站上。

杰森：所以他们只是想给我们一个惊喜？！

莉迪亚：是的。不到最后一分钟你是不会知道结果的。

杰森：哦！我敢说当他们公布结果时我会晕倒。

Unit 9 Medals and Awards

Jason: 选手

Lydia: 教练

Jason: I've heard that there's some adjustment on the given **medals**, which means, there would be less medals **eventually**.

Lydia: You mean the <u>tied</u> medals?

Jason: Tied medals? I am not so sure what those are.

Lydia: If the difference between two or more competitors is <u>no more than</u> two points on the 700 scale, then tied medals are awarded.

Jason: You mean there might be two gold medals?

Lydia: Yes sure. When there are two gold medals, there would be no **silver** medal, one or more **bronze** medal. Or, three or more gold medals, no silver medal.

Jason: Three gold medals! <u>I can't believe my ears</u>. Does that mean there would be no silver medal or bronze medal?

Lydia: When the difference between the last gold medal winner(s) and the next competitor(s) is <u>not more than</u> two points, there would be one or more bronze medals.

Jason: That sounds like a very good happy ending to me.

Lydia: But you should know that it's not a **normal** case.

Jason: So do you have any idea of what's the **maximum** number of medals that would be awarded?

Lydia: We haven't met that case before. I don't know.

Jason: But we now know that there would be maximum 3 gold medals!

▶ [VOCABULARY & PHRASE]

medal / ˈmed(ə)l /	gold medal 金牌
n. 勋章，奖章；纪念章	silver medal 银牌

bronze medal 铜牌

medal of honor 荣誉勋章

eventually / ɪˈventʃuəli /
adv. 最后，终于

tie / taɪ /
v. 打结；连接；与……成平局
n. 领带；领结；束缚；系梁；平局
tie in 使结合；使配合得当
tie with 在比赛中得分与……相同
tie up 占用
tie in with 与……一致；配合

silver / ˈsɪlvər /
n. 银；银器（尤指餐具）
adj.（有关）银的；含银的；<美>口才流利的
v. 变成银色
silver lining （不幸或失望中的）一线希望

bronze / brɑːnz /
n. 青铜；古铜色；青铜制品
adj. 青铜色的；青铜制的
v. 变成青铜色，被晒黑；镀青铜于
Bronze Age 铜器时代
bronze ware 铜器

normal / ˈnɔːrm(ə)l /
adj. 正常的；正规的；标准的
n. 正常；标准；常态
normal university 师范大学
normal operation 常规操作
normal temperature 正常体温

maximum / ˈmæksɪməm /
adj. 最高的；最多的；最大极限的
n. [数] 极大，最大限度；最大量
maximum number 最大数
maximum limit 最大限度
maximum temperature 最高温度

▶ [NOTES]

1. tie，表示"平局，打成平手"之意，多用于体育比赛或类似的竞技性活动中，比如：England tied Sweden 2-2. 或 England held a 2-2 tie with Sweden. 都表示"英国与瑞典打成 2 比 2 平"。此外，draw 也可以用来表示"平局，平手"之意，如果想表达"比赛打成了平局"，就可以说 The game resulted in a draw / tie. 或者 The game is drawn.。

2. no more than，表示"仅仅""只有"（= only），强调少；而 not more than 表示"不多于""至多"（= at most）。例如：He is no more than an ordinary English teacher. 他只不过是个普通

的英文老师。He has no more than three children. 他只有 3 个孩子。用于比较两件事物时，no more ... than 表示对两者都否定，意为"同……一样不"（=neither ... nor）；而 not more ... than 指两者虽都具有某种特征，但程度不同，意为"不如""不及"（= not so ... as）。

3. I can't believe my ears. 我不能相信我所听到的。

▶ [TRANSLATION]

第 9 单元　奖牌和奖状

杰森：　我听说对给定的奖牌进行了一些调整，这意味着最终奖牌数量会减少。

莉迪亚：你是说并列奖牌？

杰森：　并列奖牌？我不太确定那是什么。

莉迪亚：如果两个或多个参赛者成绩之间的差值在 700 分制上不超过 2 分，则会颁发并列奖牌。

杰森：　你的意思是可能会有两枚金牌？

莉迪亚：是的。当有两枚金牌时，将没有银牌，有一枚或多枚铜牌。或者，将会有三枚以上金牌，没有银牌。

杰森：　三枚金牌！我简直不敢相信自己的耳朵。这是否意味着没有银牌或铜牌？

莉迪亚：当最后一位金牌得主和下一位选手之间的差距不超过两分时，将会有一枚或多枚铜牌。

杰森：　对我来说，这听起来是一个非常美好的结局。

莉迪亚：但你应该知道这不是正常情况。

杰森：　所以你知道最多会颁发多少枚奖牌吗？

莉迪亚：我们以前没有遇到过这种情况。我不知道。

杰森：　但我们现在知道最多会有 3 枚金牌！

Unit 10 Doing an Interview

Sharon: 记者

Jason: 选手

Sharon: I <u>am with</u> the rock star from Printing Media Technology, Jason. Jason, this is your first time in the World Skills Competition, what's your favorite part about tonight?

Jason: I am so happy to be here. The best part about tonight is that I can **represent** my country to participate in the competition. And there are all the competitors from different skills. I feel so proud to walk into the **stadium** together with them.

Sharon: I know that you have won three <u>**straight**</u> medals in your **national** competition and also you come here as the **overwhelming** favorite competitors here, what's the **pressure** like?

Jason: It's a completely different challenge and I have a lot of pressure, but I'd like to put it on myself. I just want to celebrate with the rest competitors, hopefully they would be there with me. We have a large target on our back.

Sharon: I get it. And this time it's held in your hometown.

Jason: Yes. It's exciting for our country to host one of the biggest skill events in the world. I think the **buzz** around Shanghai, where obviously, I was brought up, is really exciting for everyone. It's going to be an amazing game and we're going to make it proud.

Sharon: Are you ready to bring your country a gold medal?

Jason: Yes. **Hopefully** I will. But I know that other competitors are also very **tough**. I would try my best to <u>give the best shot</u>.

Sharon: Do you have any <u>shout outs</u>?

Jason: I would like to <u>take this opportunity to</u> thank my experts, teachers, my parents and all those who had helped me. I would never <u>let you down</u>.

Sharon: Hope you can <u>get relaxed</u>.

Jason: Sure.

Sharon: Good luck.

Jason: Thank you.

[VOCABULARY & PHRASE]

represent / ˌreprɪˈzent /
v. 代表；表现；描绘；回忆；再赠送；提出异议
represent something to oneself 想象出某事物
represent for 代表；象征

stadium / ˈsteɪdɪəm /
n. 体育场；露天大型运动场
indoor stadium 室内运动场

straight / streɪt /
adj. 直的；连续的；正直的；整齐的
adv. 直接地；不断地
n. 直线
go straight 改过自新
straight up 直率地
straight line 直线
straight out 直言地

national / ˈnæʃ(ə)nəl /
adj. 国家的；国民的；民族的；国立的
n. 国民
national economy 国民经济
national standard 国家标准
national security 国家安全
National Day 国庆节

overwhelming / ˌoʊvərˈwelmɪŋ /
adj. 压倒性的；势不可挡的
overwhelming superiority 绝对优势

pressure / ˈpreʃər /
n. 压力；压迫，压强
v. 迫使；密封；使……增压
under pressure 面临压力
high pressure 高压
blood pressure 血压

buzz / bʌz /
v. 发出嗡嗡声；充满兴奋；迅速流传
n. 嗡嗡声；（愉快或兴奋等）强烈情感；活跃的气氛
buzz off 急忙离去

hopefully / ˈhoʊpfəli /
adv. 有希望地，有前途地

tough / tʌf /
adj. 艰苦的；坚强的；强壮的
n. 恶棍
v. 坚持；忍受，忍耐
adv. 强硬地，顽强地
tough guy 硬汉

▶ [NOTES]

1. be with sb.，记者在介绍"站在我身边的是……"的时候，经常用"be with sb."。
2. straight，有"连续的"意思，因此 three straight medals 意思为"连续3年拿了奖牌"。
3. give the best shot，尽最大努力。
4. shout outs，朝……喊话。
5. take the opportunity of...，利用……的机会。
6. let sb. down，让……失望。
7. get relaxed，放松。

▶ [TRANSLATION]

第10单元　接受采访

莎伦：现在站在我旁边的是来自印刷媒体技术项目的大明星杰森。杰森，这是你第一次参加世界技能大赛，今晚你最喜欢的部分是什么？

杰森：我很高兴来到这里。今晚最棒的部分是我可以代表我的国家参加比赛。和其他不同项目的选手一起走进赛场，我感觉非常棒。

莎伦：我知道你在全国比赛中连续三次获得了奖牌，而且作为最受欢迎的选手来到这里，你的压力大吗？

杰森：这是一个完全不同的挑战，我的压力很大，但我想要带着这些压力前进。我只是想和其他选手一起庆祝，希望他们能和我在一起。我们的目标都很宏大。

莎伦：我深有同感。并且这次比赛是在你的家乡。

杰森：是的。对我们国家来说，能举办世界上最大的技能赛事是很令人兴奋的。在这个我长大的地方——上海，已经感受到了热情的气氛，这对每个人来说都非常兴奋。这将是一场了不起的比赛，我们将为此感到自豪。

莎伦：你准备好为你的国家拿一块金牌了吗？

杰森：是的。希望我会。但我知道其他选手也很强。我会尽我最大的努力。

莎伦：你想对谁说点什么吗？

杰森：我想借此机会来感谢我的专家、老师、我的父母以及所有帮助过我的人。我永远不会让你们失望。

莎伦：希望你能放松好好比赛。

杰森：一定的。

莎伦：祝你好运。

杰森：谢谢。

Part 3

Oral English in Competing Days

Unit 1 Familiarization Day

（1）　大家好，我叫×××，是来自中国的选手，希望和大家一起享受这个比赛。

Hi!（正规）/ How are you doing !（熟悉后）I'm ×××, I am the competitor from China. I hope that we can enjoy this competition together.

（2）　首席（或副首席）专家，我的翻译和专家可以跟着我吗？

Chief (Vice-Chief) / Sir, could my interpreter and expert go with me?

（3）　首席（或副首席）专家，我能录像和记笔记吗？

Chief (Vice-Chief) / Sir, could I record it with my phone and take notes?

（4）　专家（或技术人员），我有多长时间熟悉这台设备？

Sir, how much time is available for me to familiarize this machine?

（5）　专家（或技术人员），这台设备我没有用过，你能给我讲解一下它的基本操作方法吗？

Sir, I haven't used this machine before. Is it possible for you to clarify the basic operation of the machine?

（6）　专家（或技术人员），这台设备有中文界面吗？怎么切换呢？

Sir, does this machine have the Chinese interface? How can I switch it to the Chinese interface?

（7）　专家（或技术人员），这台设备的×××功能怎么使用呢？

Sir, how can I use the ××× function of this machine?

（8）　专家（或技术人员），你能讲得慢一点吗？我没听太明白。

Sir, could you please speak slower? I didn't follow you.

（9） 专家（或技术人员），我能自己试一下吗？

Sir, can I try it myself?

（10） 专家（或技术人员），你觉得我这样操作正确吗？

Sir, am I doing it the right way?

（11） 专家（或技术人员），比赛期间的设备状态如何呢？

Sir, how will be the status of these machines when we are competing?

（12） 专家（或技术人员），非常感谢你的指导！

Sir, thank you very much for your guidance!

▶ [NOTES]

设备熟悉日

设备熟悉日一般安排在比赛前 2 天（C-2），是赛前准备的重要环节，也是选手在正式比赛前熟悉赛场环境，使用比赛现场设备、工具、材料的唯一机会。

一般在设备熟悉活动开始前，各项目会组织一个简短的自我介绍环节。

选手在熟悉设备期间会按照时间表安排，按序到自己所要工作的所有工位进行设备操作。熟悉时，一般各项目都会有专家（或技术人员）在各工位工作，他们的职责是保障选手和设备的安全，回答选手关于设备操作的问题。

Unit 2 Competition Day

（1） 专家，我现在可以检查一下设备吗？

Sir, can I check the machine now?

（2） 专家，能帮我把界面切换成中文吗？

Sir, could you please help me to switch it to Chinese interface?

（3） 专家，能给我一分钟让我冷静一下吗？

Sir, can I have one minute to calm down?

（4） 专家，你有什么进一步信息或说明给我的吗？

Sir, do you have any further information or instructions for me?

（5） 专家，我准备好了。

Sir, I am ready.

（6） 专家，这是我的OK样。

Sir, here is my OK sheet.

（7） 专家，这是我的校版纸。

Sir, here is my make-ready sheet.

（8） 专家，这是我的第×××张印品。

Sir, here is my No. ××× sample sheet.

（9） 专家，这是我的所有产品。

Sir, here are all my products.

（10）专家，这是我的最终色差。

Sir, here is my final Delta E.

（11）专家，这是我的测量值。

Sir, here is my measured value.

（12）专家，我可以开始清洁了吗？

Sir, can I start cleaning now?

（13）专家，我的工作完成了。

Sir, I have finished my work.

（14）专家，我需要×××（物品）。

Sir, I need ×××.

（15）专家，我需要我的翻译。

Sir, I need my interpreter.

▶ [NOTES]

比赛日

比赛时（C1 至 C4），选手在正式工作前一般可有 2~3 分钟的时间来查看工位的设备、仪器、耗材等（但不允许进行任何操作），若有试题临时改变，专家也会向选手进行简短的解释，选手要利用好这短暂的几分钟，针对不确定的内容与该工位专家进行最后确认。

工作期间，选手若有问题应尽可能用最简短的表达与专家进行沟通，因为任何在工作中的提问均不停止计时，若问题过于复杂，选手难以依靠自身英语表述清楚，或无法理解裁判的答复，或出现突发情况等，选手应尽快通过首席（或副首席）专家来呼叫自己的翻译进行协助，避免耽误过多比赛时间。

Part 4

Exercise

Unit 1 Offset Printing

1. **Please write down the corresponding English of these terms.**

（1）胶印　　　　　　　　　　　　（2）印刷机

（3）印版　　　　　　　　　　　　（4）套准

（5）纸张　　　　　　　　　　　　（6）橡皮布

（7）辊　　　　　　　　　　　　　（8）滚筒

（9）油墨　　　　　　　　　　　　（10）计数器

（11）收纸台板　　　　　　　　　　（12）飞达

（13）墨斗　　　　　　　　　　　　（14）样张

（15）弊病　　　　　　　　　　　　（16）图像

（17）印品数量　　　　　　　　　　（18）校版纸

（19）纸张规格　　　　　　　　　（20）成品规格

（21）目标密度值　　　　　　　　（22）色彩扫描

（23）压印滚筒　　　　　　　　　（24）刮墨刀

（25）印刷单元　　　　　　　　　（26）印版打孔

（27）印版弯版　　　　　　　　　（28）双张

（29）印刷速度　　　　　　　　　（30）墨键

（31）色彩（检测）区　　　　　　（32）色块

（33）专色　　　　　　　　　　　（34）预设

（35）高光涂布纸　　　　　　　　（36）光面纸

（37）翻面　　　　　　　　　　　（38）工作区域

2. **Please translate these sentences into Chinese**.

(1) This job is Printed Offset and will run 1/4 front and back.

(2) Competitors mount plates, establish registration and pre-set the press.

(3) Within the first 30 minutes no assistance will be allowed by the technicians.

(4) No overtime will be allowed.

(5) Select correct blanket and packing and install blanket and packing on press.

(6) Install ink rollers, set roller strip to 4 mm and provide the strips.

(7) Competitors will change 4 plates, establish registration and pre-set the press for high quality and productivity.

(8) Achieve color and registration and provide one signed OK sheet (No image on the back) to the judges.

(9) Give the technician your make ready waste.

(10) Set counter to "0", begin production with empty trolley.

(11) Pull one signed sample sheet for evaluation during printing at the 300th (±50) sheets printed.

(12) Immediately give the sample pull to the expert for color scanning.

(13) Competitors leave pile and plates on the press.

(14) Competitors will remove their job trolley from press and replace with an empty trolley.

(15) Wash up the units (rollers, plates, blankets and impression cylinders with automatic program).

(16) Manually clean the 4 wash-up blades in the correct and safe way.

(17) Clean the working area.

(18) Leave your workplace according to provided picture standard.

(19) Indicate to the judges when you have completed the task.

(20) Use safety equipment when touching washing chemicals.

Part 4 Exercise

Unit 2 Digital Printing

1. Please write down the corresponding English of these terms.

（1）（色彩）校正   （2）纸箱

（3）预飞   （4）拼版

（5）出血   （6）光栅图像处理器

（7）小册子   （8）封面

（9）文字   （10）印后加工

（11）模板   （12）印刷适性

（13）分辨率   （14）涂布的

（15）工作流程　　　　　　　　　（16）（电脑）桌面

（17）数字印刷　　　　　　　　　（18）数字文件

（19）可变数据　　　　　　　　　（20）裁切标记

（21）施工单　　　　　　　　　　（22）工作信息

（23）骑马订　　　　　　　　　　（24）工作标签

（25）国家代码　　　　　　　　　（26）图像分辨率

（27）裁切线　　　　　　　　　　（28）热文件夹

（29）页面列表　　　　　　　　　（30）色彩管理

2. Please translate these sentences into Chinese.

(1) This job is Printed Digital.

(2) Load papers into press and set up.

(3) Ensure that you have the right trim size and bleeds (3 mm).

(4) Give required quantity to judges.

(5) Digital files are in the folder "DIGITAL 1" on the desktop of computer.

(6) Preflight the files in "DIGITAL 2" folder and take notes.

(7) Competitors will perform calibration procedure.

(8) Take notes of the font, size and position information.

(9) Competitors will load specified paper into press and set up.

(10) You will need to perform a front to back registration test.

(11) Ask technician to fix machine jams if required.

(12) Competitors will open a PDF file provided by the judges.

(13) The competitor is to preflight the file to assure printability.

(14) Preflight the digital file to check following specifications for the usual faults.

(15) Create a single new file with only the files that are acceptable specifications.

(16) Register the front to back images to within 0.6 mm in each corner of the press sheet.

(17) Give the judges your final "OK" sheet.

(18) Print 10 copies of your job.

Unit 3 Additional Task

1. Please write down the corresponding English of these terms.

（1）维护保养　　　　　　　　　　　（2）设备

（3）千分尺　　　　　　　　　　　　（4）（橡皮布）包衬

（5）划痕　　　　　　　　　　　　　（6）墨皮

（7）起脏　　　　　　　　　　　　　（8）不干胶

（9）匹配　　　　　　　　　　　　　（10）工具

（11）计算器　　　　　　　　　　　　（12）容器

（13）手动地　　　　　　　　　　　　（14）模拟

（15）水辊　　　　　　　　　　　　（16）墨辊

（17）墨辊压痕量规　　　　　　　　（18）着墨辊

（19）弊病鉴别　　　　　　　　　　（20）放大镜

（21）水墨平衡　　　　　　　　　　（22）橡皮布塌陷

（23）非图像区域　　　　　　　　　（24）参考样张

（25）故障查找　　　　　　　　　　（26）油墨秤

（27）潘通色卡　　　　　　　　　　（28）墨刀

（29）色差　　　　　　　　　　　　（30）切纸机

（31）裁切顺序　　　　　　　　　　　（32）模块程序

2. **Please translate these sentences into Chinese**.

(1) Safely install and adjust rollers, plates, blankets on the Offset Printing Press.

(2) Build the blanket and packing to a thickness of 3 mm.

(3) Mount the blanket with packing on press.

(4) Install #4 ink form roller into press.

(5) Set roller stripe on all four ink form rollers and dampening roller.

(6) Circle all defects you find on each of the press sheets.

Part 4 Exercise

(7) Identify common offset printed defects.

(8) The competitor with the fastest time will receive full time marks.

(9) Follow proper procedures for blanket installation.

(10) Place the Pantone ink into the ink fountain.

(11) Measure the Lab values of your ink mix.

(12) Remove the ink from the fountain.

(13) Mix an appropriate quantity of Pantone ink.

(14) When you have reached the Lab you like, call the judges.

(15) All mixed ink must be placed in container.

(16) No ink waste allowed.

(17) Competitors with the most color spots with the lowest Delta E will receive full marks.

(18) Indicate cutting with a drawn line on one printed sheet.

(19) Number the lines to indicate the order of cutting.

(20) Program cutter and cut the tent cards.

(21) Keep finished product in sequence order.

(22) Errors come from not following standard press operating procedures for make ready.

(23) The system will ask you if you want to exit as soon as the exercise is completely finished "printing" good copies.

(24) If you attempt to restart an exercise during the exercise, you will receive zero marks for this exercise.

(25) Target Lab values are supplied in eXact.

(26) Perform cutting in the safest possible manner.

Appendix 1

Expanded Reading

WORLDSKILLS STANDARDS SPECIFICATION

	SECTION	RELATIVE IMPORTANCE /%
1	**Work organization and management**	10
	The individual needs to know and understand: • The types of equipment used to produce printed materials and to create finished printed products • New technologies used in printing • Current legislation and best practice relating to health and safety procedures in the workplace and specifically relating to specialist equipment and print factors • The uses of protective equipment and materials • The importance on maintaining cleanliness and order in the working environment • The handling of chemicals used in the print industry and how waste materials should be disposed • The importance of effective communication skills and team work • Recognized international standards, for example ISO, GRACoL and Pantone	
	The individual shall be able to: • Use all equipment correctly according to manufacturers' instructions • Consistently apply and promote health and safety procedures in the workplace, especially relating to specialist equipment and print factors • Effectively use protective equipment and materials	

(continued)

	SECTION	RELATIVE IMPORTANCE /%
1	**Work organization and management**	10
	• Maintain a clean and ordered working environment • Handle all chemicals and dangerous materials safely and in accordance with instructions • Dispose of waste materials safely and consistently to maintain a safe and sustainable environment • Select equipment appropriate for the planned task • Use, handle, store and maintain print factors such as ink, paper and mechanical and digital equipment • Proactively maintain continuous professional development in order to keep up to date with new technologies and trends in the printing industry • Recognize a suitable file for digital printing (preflight) • Assure the quality and check all work to verify and adjust details in the quality of the printed work and to ensure that it meets customer's expectations and high standards • Communicate effectively with team members and other colleagues in the workplace to ensure a good and productive working environment • Discuss client's requirements, and provide expert advice and guidance on printing technology, its possibilities and limitations • Try best to avoid unnecessary waste	

(continued)

SECTION		RELATIVE IMPORTANCE /%
2	**Planning and preparation**	20
	The individual needs to know and understand: • The characteristics of the Offset Printing, toner-based and ink-jet digital materials • The characteristics, uses and interaction of papers, inks, dampening solution, toners and proofing materials • Appropriate chemicals needed for the planned print job • The theory of colors • How to read, understand and analyse a customer's brief	
	The individual shall be able to: • Read, understand and interpret a print job brief • Explain the content of the brief to colleagues and plan work for self and others accordingly • Mix custom ink colors to meet customers' specifications • Select and prepare the appropriate printing equipment for the planned job • Programme machinery for correct number of copies, paper size, color, quality, etc. • Perform make-ready operation and adjustment on multi-color sheet-fed offset press, either with or without remote control consoles • Select and prepare the appropriate print factors, paper, ink etc. for the planned job • Interpret the color imprint on the printed sheet during make-ready and production	

(continued)

	SECTION	RELATIVE IMPORTANCE /%
2	**Planning and preparation**	**20**
	• Translate the interpretation of the color imprint into appropriate action on the press • Use digital printing press RIP (Raster Image Processor) software for file set-up operations, like checking/creating, imposition and color management, and use Variable Data Printing (VDP). • Load paper sheets and fill ink ducts • Adjust the feeder, sheet transfer and delivery • Mount offset printing plates • Adjust offset printing pressure • Mix necessary ink colors • Adjust the color register	
3	**Press run**	**20**
	The individual needs to know and understand: • Different types of press and their uses and characteristics • Developing technology that supports the printing process	
	The individual shall be able to: • Print a specific amount of printed products on the sheet-fed offset press according to the quality and technical criteria set, industry standard and standard required by the customer • Use presses with either semi-automatic or automatic plate mounting • Use variable data software for digital printing	

(continued)

SECTION		RELATIVE IMPORTANCE /%
4	**Quality control, adjustments, and troubleshooting**	40
	The individual needs to know and understand: • Different types of specialist measuring equipment used in printing • How to interpret measuring results • The importance of ensuring that the print job is of a high standard and meets the customers' needs and expectations • The financial and virtual time reporting functions of sheet-fed press simulation programmes • Implications of faulty machinery or set-up in terms of loss of quality, time and money • Maintenance routines for printing equipment • The importance of following manufacturers' instructions	
	The individual shall be able to: • Monitor the printing process, ensuring that the work is reaching the expected standard • Adjust settings and programming to maintain quality and rectify discrepancies with the specification • Operate measuring and quality control devices • Use different measuring devices like densitometer, spectrophotometer, micrometer, calliper, pH, conductivity, etc. • Produce OK sheet for customers' confirmation and approval • Save individual sheets as prescribed through the print run to quality assure against the original • Compare proof prints to specified targets and make necessary adjustments	

(continued)

	SECTION	RELATIVE IMPORTANCE /%
4	**Quality control, adjustments, and troubleshooting**	40
	• Produce print jobs to a specified numerical density and/or Lab color space target • Maintain the correct color registration • Solve problems in the sheet-fed press simulation programme • Perform maintenance and basic repairs on offset presses and finishing equipment • Resolve paper feed problems	
5	**Finishing**	5
	The individual needs to know and understand: • Various processes that may be applied to printed work to finish the product such as folding, cutting and binding	
	The individual shall be able to: • Prepare a cutting plan • Finish printed work by trimming it with a paper cutter to specified dimensions • Programme and use a programmed paper cutter to cut paper to specified dimensions • Operate a digital in line stitcher or perfect binding machine to produce bound printed works	
6	**Clean-Up**	5
	The individual needs to know and understand: • The advantages of working in a clean and ordered environment	

（continued）

SECTION		RELATIVE IMPORTANCE /%
6	**Clean-Up**	5
	The individual shall be able to: • Clean the equipment and premises after the offset, digital and finishing printing process • Complete cleaning efficiently, effectively and within prescribed timescales • Ensure that cleaning is completed to recognized standards • Set back adjustments of the printing equipment to zero	
	Total	100

《世界技能标准规范》

部分		重要性/%
1	工作组织和管理	10
	选手需知道和了解： • 制备印刷材料、生产印刷产品所使用的设备类型 • 印刷中的新技术 • 工作区域内，健康安全规则涉及的现行法规及最佳实践方法，特别是专业设备和印刷要素涉及的现行法规及最佳实践方法 • 保护装置和材料的使用 • 保持工作区域内清洁、有序的重要性 • 印刷业中化学用品的使用及废料处理 • 有效的沟通技巧及团队合作的重要性 • 掌握国际标准，如 ISO、GRACoL 和 Pantone 等	
	选手必须能够： • 根据制造商的说明正确操作所有设备 • 在工作区域内持续运用和提倡健康安全规则，特别是专业设备和印刷工艺要素方面 • 有效使用保护装备和材料 • 保持工作环境清洁、有序 • 按照说明，安全使用化学用品和危险材料 • 安全地处置废料，始终维持安全且可持续发展环境 • 选择适合计划任务的设备 • 使用、处理、存储和维护保养印刷要素，如油墨、纸张、机械和数字设备 • 主动保持持续的专业发展，以跟上印刷行业的新技术和新趋势 • 保证数字印刷（预飞）的适当文件（格式） • 确保质量，检查各项工作，以验证和调整印刷工作的质量细节，从而确保其符合客户的预期和高标准	

（续表）

	部分	重要性/%
1	**工作组织和管理**	10
	• 在工作区域内与团队成员和其他同事有效沟通，维持良好、高效的工作环境 • 探讨客户需求，就所采用的印刷技术、能力和局限性提供专业性建议和指导 • 工作中尽量避免不必要的浪费	
2	**计划和准备**	20
	选手需知道和了解： • 胶印、基于墨粉和喷墨的数字印刷材料的特点 • 纸张、油墨、润版液、墨粉和打样材料的特点、使用及相互作用关系 • 完成印刷工作所需的合适化学用品 • 色彩理论 • 如何阅读、领会和分析客户说明	
	选手必须能够： • 阅读、领会和诠释印刷工作说明 • 向同事解释说明和计划内容，并为自己和他人制订工作计划 • 配置客户指定的油墨颜色，满足客户的需求 • 根据工作计划选择和准备合适的印刷设备 • 为准确的成品数量、纸张尺寸、颜色、质量等设定程序 • 无论使用或不使用远程控制台，对多色单张纸胶印机进行预印操作和调整 • 根据工作计划选择和准备合适的印刷要素，如纸张、油墨等。 • 在预印和生产期间，解释印张上的印色 • 将印色的解读转化为恰当的印刷操作 • 使用数字印刷机 RIP（光栅处理器）软件来设置文件，如检查/创建、拼版和色彩管理，并使用可变数据印刷（VDP）	

（续表）

	部分	重要性/%
2	计划和准备	20
	• 装纸和注墨 • 调整飞达、递纸和收纸装置 • 安装印版 • 调整胶印压力 • 调配所需油墨色彩 • 调整色彩套准	
3	印刷	20
	选手需知道和了解： • 印刷机的不同类型及其使用方法和特点 • 支持印刷工作流程的发展技术	
	选手必须能够： • 根据质量和技术规则、工业标准和客户需求，在单张胶印机上印刷指定数量的印刷产品 • 使用半自动装版和全自动装版的印刷机 • 使用可变数据软件进行数字印刷	
4	质量控制、调整、故障排除	40
	选手需知道和了解： • 印刷过程中使用的不同类型的专业测量设备 • 如何解读测量结果 • 确保高质量印刷工作，以满足客户需求和期望的重要性 • 单张胶印机模拟程序的花费和虚拟时间报告功能 • 机械或装置故障对质量低下、费用和时间增多的影响 • 印刷设备的维护 • 遵守制造商说明的重要性	

（续表）

	部分	重要性/%
4	质量控制、调整、故障排除	40
	选手必须能够： • 监控印刷过程，确保工作达到预期标准 • 调整设施与程序以保证质量，并纠正与规范的差异 • 操作测量和质量控制装置 • 使用不同的测量设备，如密度计、分光光度计、千分尺、游标卡钳、pH 值测试仪、导电率测试仪等 • 制作 OK 样张供客户确认和批准 • 印刷过程中按规定保存个别印张，以确保质量与原样张一致 • 将印刷样张与指定的目标进行比较，并进行必要的调整 • 按指定密度和/或 Lab 值完成印刷工作 • 保持正确的颜色套准 • 解决单张胶印机模拟系统出现的问题 • 对胶印机和印后加工设备进行保养和基础维修 • 解决输纸问题	
5	印后加工	5
	选手需知道和了解： • 可应用于印刷品以完成印刷工作的各工序，例如折页、裁切和装订	
	选手必须能够： • 制订裁切方案 • 通过切纸机将印刷品裁切到指定尺寸以完成印刷工作 • 编程并使用可编程的切纸机将纸张裁切到指定尺寸 • 使用数字印刷联线骑马订机或胶装机，进行装订工作	
6	清扫	5
	选手需知道和了解： • 在整洁、有序的环境中工作的优势	

（续表）

部分		重要性/%
6	清扫	5
	选手必须能够： • 在完成胶印、数字印刷和印后加工后，清洁设备和场地 • 在规定的时间内高效及有效地完成清洁工作 • 确保清洁达到公认标准 • 将印刷设备调整状态归零	
	总分	100

Appendix 2

Vocabulary & Phrase

A

abbreviation / əˌbriːviˈeɪʃn / *n.* 缩写；缩写词
English abbreviation 英语缩略语；abbreviation index 缩略词对照表；unit abbreviation 单位缩写

access / ˈækses / *n.* 通道；进入；入口 *v.* 接近，使用；访问，存取（电脑文档）*adj.* <美>（电视节目或时间等）对外公开的
access control 访问控制

accommodation / əˌkɑːməˈdeɪʃ(ə)n / *n.* <美>住处，膳宿；调节；和解
accommodation cost 住宿费用

accredit / əˈkredɪt / *v.* 把……归于，归因于；委派；信任，正式认可；授权
accredited *adj.* 公认的；可信任的

accreditation / əˌkredɪˈteɪʃn / *n.* 委派；信赖；鉴定合格
accreditation body 认可机构；accreditation system 认可体系

acronym / ˈækrənɪm / *n.* 首字母缩略词
acronym list 简写对照表

activity / ækˈtɪvəti / *n.* 活动；行动；活跃
economic activity 经济活动；catalytic activity 催化活性；physical activity 体育活动；biological activity 生物活性

actually / ˈæktʃuəli / *adv.* 实际上；事实上
illuminations actually measured 实测照度

adapt / əˈdæpt / *v.*（使）适应；改编

adapt to 适合；改编；使适应；adapt oneself to 使自己适应于；adapt for 使适合于；调剂

additional / əˈdɪʃən(ə)l / *adj.* 附加的，额外的
additional tax 附加税

additional mark 附加分

additional task 附加任务

adhesive / ədˈhiːsɪv / *adj.* 黏着的，有黏性的 *n.* 黏合剂，胶水
gloss adhesive stock 高光不干胶

adhesive paper 不干胶

adjust / əˈdʒʌst / *v.* 调整；整理（衣服）；习惯；评估（损失、损害）
adjust to 适应

adjustment / əˈdʒʌstmənt / *n.* 调整，（行为、思想的）调节；调节器
seasonal adjustment 季节性调整；structural adjustment 结构调整；price adjustment 价格调整；speed adjustment 速度调节

airport / ˈeəpɔːrt / *n.* 机场；航空站（港）
airport terminal 航站楼

announce / əˈnaʊns / *v.* 宣布；述说；预示；播报；通报……的到来；宣布竞选公职
announce results 公布成绩；announce the opening of 宣布……的召开；announce to sb. 通知某人

align / əˈlaɪn / *v.* 使一致；公开支持；（使）排成一条直线
align with 重合；text-align 文本对齐

allotment / əˈlɑːtmənt / n. 分配；配额
time allotment 时间分配

applicable / ˈæplɪkəbl / adj. 适用的，适当的
not applicable 不适用

appropriate / əˈprouprɪeɪt / adj. 合适的，相称的 v. 拨出（款项）；私占，挪用
appropriate penalty 适当处分；be appropriate for 合适于

arena / əˈriːnə / n. 竞技场
Allianz Arena 安联竞技场

arrangement / əˈreɪndʒmənt / n. 布置；整理；准备；安排
tour arrangement 旅行安排

aspect / ˈæspekt / n. 方面；方向；形势；外貌
aspect ratio 纵横比；aspect angle 视界角；environmental aspect 环境因素

assemble / əˈsemb(ə)l / v. 集合，（使）聚集；装配；收集
assemble line 装配线

assess / əˈses / v. 评估；征税，处以罚金
assess risks 资产风险；风险评估

assessment / əˈsesmənt / n. 评判，评价
assessment method 考核方法；market assessment 市场评估

attachment / əˈtætʃmənt / n. （邮件）附件；连接物；附属物；依恋；（临时的）委派；信念；扣押
wash-up attachment 刮墨槽

attempt / əˈtempt / v. 尝试，努力 n. 试图；企图杀害；（运动员创造纪录的）尝试，冲击
attempt to 尝试

automatic / ˌɔːtəˈmætɪk / adj. 自动的
automatic control 自动控制

available / əˈveɪləb(ə)l / adj. 可获得的；可购得的
available material 可用材料
available time for the total task 工作总时间

award / əˈwɔːrd / n. 奖；（收入的）增加；（赔偿）裁定额；（毕业证书等的）授予；奖学金 v. 把（某物）授予（某人）；把（合同、佣金）给（人、组织）
Hugo Award 雨果奖；literary award 文学奖

B

badge / bædʒ / n. 徽章；证章；标记 v. 授给……徽章
arm badge 袖章；name badge 姓名牌；detective boys badge 侦探徽章

balance / ˈbæləns / n. 天平；平衡；账户余额 v. 保持平衡
ink and water balance 水墨平衡

basically / ˈbeɪsɪklɪ / adv. 主要地，基本上
basically stable 基本稳定

bench / bentʃ / n. 长条形工作台；长凳；替补席
work bench 工作台

benchmark / ˈbentʃmɑːrk / n. 基准；标准检查程序

benchmark interest rate 基准利率；benchmark system 基准系统；benchmark test 基准测试

bend / bend / v. 弯（版）

bend in 把……往里弯；bend over 俯身；折转

beside / bɪˈsaɪd / prep. 在……旁边；与……相比 adv. 在附近；况且，此外

beside the point 离题

blade / bleɪd / n. 刀片，刀刃

wash-up blade 刮墨刀

blank / blæŋk / adj. 空白的；毫无表情的；单调的；断然的 n. 空白处；（头脑或记忆）一片空白 v. 成为空白

blank press 未设置的印刷机

blanket / ˈblæŋkɪt / n. 橡皮布；毯子；覆盖层；气氛 v. 覆盖；消除（声音）adj. 总括的，全面的

blanket order 总订单

bleed / bliːd / n. 出血，流血；泄出（液体，气体）v. 出血，流血；给……放血；散开

bleed out 渗出；air bleed 排气阀

block / blɑːk / n. 大块；街区；大楼；障碍物，阻碍 v. 阻塞；遮住（视线）；封锁，挡住（去路）；阻挠；阻截；对……进行块操作

block programming 模块程序

booklet / ˈbʊklət / n. 小册子

residence booklet 户口簿

Brazilian / brəˈzɪliən / adj. 巴西的；巴西人的 n. 巴西人

Brazilian market 巴西市场

breach / briːtʃ / n. 违背，违反；缺口 v. 违反；破坏；打破

breach of contract 违约；data breach 数据泄露

bronze / brɑːnz / n. 青铜；古铜色；青铜制品 adj. 青铜色的；青铜制的 v. 变成青铜色，被晒黑；镀青铜于

Bronze Age 铜器时代；bronze ware 铜器

buzz / bʌz / v. 发出嗡嗡声；充满兴奋；迅速流传 n. 嗡嗡声；（愉快或兴奋等）强烈情感；活跃的气氛

buzz off 急忙离去

C

calculate / ˈkælkjuleɪt / v. 计算；以为；预测，推测

calculate area 计算面积

calculator / ˈkælkjuleɪtər / n. 计算器

pocket calculator 袖珍计算器

calibration / ˌkælɪˈbreɪʃn / n. （色彩）校正；（测量器具上的）刻度

calibration procedure （色彩）校正流程；screen calibration 屏幕校准

cart / kɑːrt / *n.* 收纸台板；马车；小型机动车；购物车，手推车 *v.* 用车装运；搬运；挟持，带走
 hand cart 手推车
ceremony / 'serəmoʊni / *n.* 典礼，仪式
 opening ceremony 开幕式；closing ceremony 闭幕式；award ceremony 颁奖典礼
chest / tʃest / *n.* 胸，胸部；衣柜；箱子；＜英＞金库
 on one's chest [口语] 心中有事，郁积在心，有心事
cleanliness / 'klenlməs / *n.* 清洁
 surface cleanliness 表面洁净度
cluster / 'klʌstər / *n.* 群；簇；丛；串 *v.* 群聚；丛生
 a cluster of 一群；cluster sampling 分组取样；star cluster 星团
coated / 'koʊtɪd / *adj.* 涂布的；覆盖着的 *v.* 外面覆盖（coat 的过去分词）
 coated paper 涂布纸
coated one sided 单面涂布
color correct 色彩修正
color management 色彩管理
color zones 色彩（检测）区
common offset printed defects 常见胶印弊病
compatriot / kəm'peɪtrɪət / *n.* 同胞；同国人 *adj.* 同胞的；同国的
 overseas compatriot 海外同胞
comprise / kəm'praɪz / *v.* 包含；由……组成

be comprised of 由……组成
consideration / kənˌsɪdə'reɪʃ(ə)n / *n.* 考虑；原因；关心；报酬
 under consideration 在考虑中
consistency / kən'sɪstənsi / *n.* 一致性；黏稠度
 color consistency 色彩稳定性
container / kən'teɪnər / *n.* 容器
 waste container 垃圾桶
content / 'kɑːntent / *n.*（书、文章、演讲、电影等的）内容；目录；所含物 *adj.* 满足的 *v.* 使满意，使满足
 free content 免费内容；water content 含水量；content provider 内容提供者
counter / 'kaʊntər / *n.* 计数器 *adj.* 反面的，对立的
 loop counter 循环计数器
country code 国家代码
cover / 'kʌvər / *n.* 封面；盖子；掩护；被子 *v.* 覆盖；涉及；报道；翻唱；行走（一段路程）
 hard cover 精装；soft cover 平装
craftsman / 'kræftsmən / *n.* 工匠；手艺人；技工
 skilful craftsman 能工巧匠
creep / kriːp / *v.* 爬行；蔓延；慢慢地移动；（由于恐惧、厌恶等）起鸡皮疙瘩 *n.* 爬行；＜非正式＞毛骨悚然的感觉
 creep in 悄悄混进
customer / 'kʌstəmər / *n.* 顾客

customer service 客户服务

cutting / ˈkʌtɪŋ / *n.* 裁切；剪报
cutting plan 裁切计划

cyan /ˈsaɪən; ˈsaɪæn / *n.* 青色（用于印刷） *adj.* 青色的

cylinder / ˈsɪlɪndər / *n.* （印刷机）滚筒；圆柱体，圆筒；汽缸
gas cylinder 煤气罐

D

damaged sheet 纸张损坏

dampening roller 水辊

deduct / dɪˈdʌkt / *v.* 扣除，减去
deduct money 扣钱；扣款；deduct from 从……中减去；deduct wages 扣工资

deduction / dɪˈdʌkʃn / *n.* 扣除，减除；推论；减除额
salary deduction 罚薪；tax deduction 减免税款

defect / ˈdiːfekt / *n.* 弊病；缺陷，毛病 *v.* 背叛，叛变
zero defect 零缺陷；surface defect 表面损坏

delegate / ˈdelɪɡeɪt / *n.* 代表，会议代表；委员会成员 *v.* 授（权），把……委托给他人；委派……为代表，任命
technical delegate 技术代表；business delegate 业务代表；chief delegate 首席代表

deliver / dɪˈlɪvər / *v.* 输送（纸张）；投递；履行；交付；发表；分娩；解救
deliver letters 送信；deliver goods 交货；cash on deliver 货到付款；deliver a speech 发表演讲

delivery / dɪˈlɪvəri / *n.* 收纸；递送，投递；分娩
time of delivery 交货时间

Delta E 色差

density / ˈdensəti / *n.* 密度；稠密，密集
energy density 能量密度

deserve / dɪˈzɜːrv / *v.* 应受，应得
deserve to do sth. 值得干某事；deserve one's reputation 名不虚传

desktop /ˈdesktɑːp / *n.* （电脑）桌面；台式机
clear the desktop 清理桌面

detection / dɪˈtekʃ(ə)n / *n.* 鉴别；察觉；侦破（案件）
defect detection 弊病鉴别

digital / ˈdɪdʒɪt(ə)l / *adj.* 数字的
digital printing 数字印刷；digital file 数字文件；digital printing press 数字印刷机

dimension / daɪˈmenʃn; dɪˈmenʃn / *n.* 尺寸；（空间的）维度；规模；方面 *v.* 切削（或制作）成特定尺寸；标出尺寸
three dimension 三维

display / dɪˈspleɪ / *n.* （计算机屏幕上的）显示；炫耀 *v.* （计算机）显示；表现；陈

列 *adj.* 展览的；陈列用的
LCD display 液晶显示器；on display 展出

dispute / dɪˈspjuːt / *n.* 辩论；争吵；意见不同 *v.* 辩论；对……进行质疑；争夺；抵抗（进攻）
dispute resolution 调解纠纷；in dispute 在争论中；labor dispute 劳动纠纷

double sheets 双张

drawer / drɔːr / *n.* 纸箱；抽屉
open the drawer 打开抽屉

E

edge of sheet 纸张边缘

edit / ˈedɪt / *v.* 编辑，校订；编选；剪辑；主编 *n.* 编辑，校订，剪辑
text edit 文本编辑

efficiency / ɪˈfɪʃ(ə)nsɪ / *n.* 效率；（机器的）功率
economic efficiency 经济效率；high efficiency 高效率

electro-car / ɪˈlektroʊ- kɑːr / *n.* 电动汽车

empty trolley 空的收纸台板

environmentally friendly 环保

equipment / ɪˈkwɪpmənt / *n.* 设备；（做某事应具备的）素质，才能
peripheral equipment 外部设备；medical equipment 医疗器械

equipment available 可用设备

establish / ɪˈstæblɪʃ / *v.* 建立；证实
establish its own 自立门户

eventually / ɪˈventʃuəlɪ / *adv.* 最后，终于

exchange / ɪksˈtʃeɪndʒ / *n.* 交换；交流 *v.* 交换；交易
exchange rate 汇率；exchange student 交流生；stock exchange 证券交易所；exchange for 交换；foreign exchange 外汇

excursion / ɪkˈskɜːrʒ(ə)n / *n.* 短途旅行，远足
excursion train 旅行列车；excursion boat 游艇

extra / ˈekstrə / *adj.* 额外的；另外收费的；特大的 *n.* 额外的事物；另外收费的事物；（非击球所得的）附加分 *adv.* 额外；特别地，格外地 *pron.* 额外的东西（尤其指钱财）
extra time 额外时间；加赛时间；extra money 额外的钱；extra work 加班；额外工作；extra large 加大码的；extra charge 附加费

F

faint / feɪnt / *adj.* 模糊的；头晕的 *v.* 昏倒；变得微弱 *n.* 昏厥，昏倒
faint scent 清香；faint with 由于……原因而衰弱

fair / fer / *adj.* 公平的；美丽的，（肤色）白皙的 *adv.* 公平地；直接地；清楚地 *v.* 转晴 *n.* 展览会

169

vanity fair 名利场；fair use 合理使用；fair play 公平竞赛

familiarization / fəˌmɪliəraɪˈzeɪʃn / *n.* 熟悉，精通；亲密
familiarization training 熟悉训练；familiarization visit 了解工作情况的访问；familiarization trip 熟悉之旅

fault / fɔːlt / *n.* 故障；错误 *v.* 挑剔，指责
page fault 页面错误

fault finding 故障查找

fault stop 故障（造成的）停止

faultless / ˈfɔːltləs / *adj.* 完美的；无缺点的
faultless operation 无事故运行；faultless responsibility 无过失责任制

feeder / ˈfiːdər / *n.* 飞达；支线；喂食器；奶瓶；饲养员
feeder stop 飞达停止

feeding / ˈfiːdɪŋ / *n.* 输纸；饲养
feeding system 控食系统

fiber / ˈfaɪbər / *n.* 纤维（等于 fibre）
paper fibers 纸张纤维；glass fiber 玻璃纤维

finished copies 印品

finished paper stack 收纸堆

finished sheets （印刷）完成的印张

finished size 成品规格

finished size before trimming 裁切前尺寸

finishing / ˈfɪnɪʃɪŋ / *n.* 印后加工；得分的表现和技巧 *v.* 完成；结束；用完，吃光（finish 的现在分词）
surface finishing 表面修整

first printed side 第一印刷面

fix / fɪks / *v.* 维修；安排；处理（问题等）；固定；盯住；（注意力）集中在 *n.* （尤指简单、暂时的）解决方法
fix machine jams 排除设备故障

fold / fəʊld / *v.* 折叠；可折叠 *n.* 褶层；褶痕
fold increase 成倍增加；fold line 折线；折纹；返折线

font / fɑːnt / *n.* 字体；根源，来源
font size 字体大小；embed font 内嵌字体

forbidden / fərˈbɪdn / *v.* 禁止；妨碍；禁止……进入；阻止（forbid 的过去分词）*adj.* 不被允许的，被禁止的；禁戒的
The Forbidden City 紫禁城；故宫

format / ˈfɔːrmæt / *n.* （书或杂志的）版式 *v.* 为……编排格式；格式化
file format 文件格式；output format 输出格式

form roller 着墨辊

formula / ˈfɔːrmjələ / *n.* 配方，处方；方案；公式
ink formula （专色）油墨配方

fountain / ˈfaʊnt(ə)n / *n.* （胶印机）墨斗；喷泉；天然泉
musical fountain 音乐喷泉

four color press 四色印刷机

front and back 正面和反面

front-to-back registration 正反面套准

full speed （印刷机）最大印刷速度

G

gloss / glɑːs / *n.* 光彩；光泽涂料；注释
　　gloss coated 高光涂布；gloss oil 上光油

grab / græb / *v.* 攫取；霸占；将……深深吸引 *n.* 攫取；夺取之物
　　up for grabs 大家有份；grab a bite 吃点东西；grab at 抓住，抓取

grain / greɪn / *n.* 纹理；谷物；颗粒；（表面的）质地 *v.* 使表面（或纹理）粗糙；成粒状
　　add grain 添加杂点

gsm 克 / 平方米，gram per square metre 的缩写，与 g/m² 意思相同

H

Heidelberg procedure 海德堡（工作）流程

Heidelberg Speedmaster 海德堡速霸

heritage / ˈherɪtɪdʒ / *n.* 传统；遗产
　　cultural heritage 文化遗产；World Heritage Site 世界遗产保护区

hickey / ˈhɪki / *n.* 墨皮；器械；唇印

hold / hoʊld / *v.* 手持；容纳；举办；持有观点；保存（如在计算机中）
　　X-Rite eXact hand held 爱色丽 eXact 手持式分光光度仪
　　hold on 坚持；稍等；hold up 举起；hold off 推迟

hopefully / ˈhoʊpfəli / *adv.* 有希望地，有前途地

host / hoʊst / *n.* 主人；主持人 *v.* 主持；当主人招待
　　host country 东道国；host family 寄宿家庭；host city 主办城市

hot folder 热文件夹

I

identity / aɪˈdentəti / *n.* 身份，本体；个性，特性；同一性，一致；恒等运算，恒等式
　　an identity of interests 利益共同体

image / ˈɪmɪdʒ / *n.* 图像；形象，印象 *v.* 作……的像，描绘……的形象
　　image sensor 图像传感器；digital image 数字图像；virtual image 虚拟图像

imagery / ˈɪmɪdʒəri / *n.*（艺术作品中的）像；意象；比喻；形象化
　　satellite imagery 卫星图；mental imagery 心理意象；visual imagery 视觉表象

immediately / ɪˈmiːdiətli / *adv.* 立即，立刻；直接地 *conj.* <英> 一……就
　　immediately following 紧跟着；reply immediately 秒回

import / ˈɪmpɔːrt / *v.* 输入；导入（计算机）

n. 进口；输入；重要性

import duty 进口税；import licence 进口许口证

imposition / ˌɪmpəˈzɪʃ(ə)n / *n.* 拼版

turn imposition 翻面

impression / ɪmˈpreʃ(ə)n / *n.* 压印，印记；印次

impression cylinder 压印滚筒

imprint / ɪmˈprɪnt / *v.* 印刷 *n.* 印记；痕迹

raindrop imprint 雨痕

indicate / ˈɪndɪkeɪt / *v.* 表明；指出；预示；象征

indicate right 打右转向灯；evidence indicate 迹象；预示；evidence-indicate 表明

information / ˌɪnfərˈmeɪʃ(ə)n / *n.* 信息；消息，资料，情报；信息台

information system 信息系统；information technology 信息技术；basic information 基本信息

ink fountain 墨斗

ink key （胶印机）墨键

ink knives 墨刀

ink waste 浪费油墨

inspection / ɪnˈspekʃn / *n.* 检查；视察

inspection sheet 校准印张；inspection certificate 检验证书

inspire / ɪnˈspaɪər / *v.* 激发（想法）；鼓舞；启示

inspire a generation 激励一代人；inspire sb. to do sth. 激励某人做某事

install / ɪnˈstɔːl / *v.* 安装；正式任命；安顿

easy to install 安装灵活

installation / ˌɪnstəˈleɪʃ(ə)n / *n.* 安装，装置；就职

installation manual 安装手册

intent / ɪnˈtent / *n.* 目的；故意 *adj.* 专注的；坚决的；急切的

output intent profile 输出目标特征文件

intention / ɪnˈtenʃ(ə)n / *n.* 意图；目的

purchase intention 购买意愿；original intention 初衷

interpreter / ɪnˈtɜːrprətər / *n.* 口译者

simultaneous interpreter 同声传译员

involve / ɪnˈvɑːlv / *v.* 包含；牵涉；使陷于；潜心于

involve in 牵涉

isolate / ˈaɪsəleɪt / *v.* 隔离；分离；区别看待 *adj.* 孤独的，孤立的 *n.* 被隔离的人（或物）

isolated workspace 隔离的工作区域

J

job information 工作信息

job ticket 施工单

job trolley 收纸台板

judgement / ˈdʒʌdʒmənt / *n.* 评价（评分）；意见；判断力；审判

subjective judgement 主观判断

L

label /ˈleɪbl/ *n.* 标签；称号；商标；唱片公司；（计算机）标记 *v.* 贴标签；把……不公正地称为
job label 工作标签；logo label 商标

lanyard /ˈlænjərd/ *n.* （用于悬挂哨子、证件等的）挂带
personal retention lanyard 个人保护系索

laundry /ˈlɔːndrɪ/ *n.* 洗衣店；洗衣房
laundry detergent 洗衣粉；laundry service 洗衣服务；laundry bag 洗衣袋；laundry room 洗衣房

layout /ˈleɪaʊt/ *n.* 拼版；设计；版面设计
create layout 进行拼版；page layout 页面布局

live job 真实的工作

load /loʊd/ *v.* 装入（纸张）；上（子弹）；载入（计算机程序）*n.* 负载，重荷
full load 满载

locate /ˈloʊkeɪt/ *v.* 确定……的位置，探明
locate the fault 找出问题

loop /luːp/ *n.* 环形；（程序中）一套重复的指令；回路；<英>（铁道或公路的）环线 *v.* 使绕成圈；执行计算机指令
event loop 事件循环

M

magenta /məˈdʒentə/ *n.* 品红；洋红 *adj.* 品红色的；洋红色的

magnify /ˈmæɡnɪfaɪ/ *v.* 放大；夸张；使（问题）加重
loop magnifying glass 环形放大镜

maintenance /ˈmeɪntənəns/ *n.* 维护，保养；（依法应承担的）抚养费
maintenance area 维修区

make note 做笔记

manager /ˈmænɪdʒər/ *n.* （公司、部门等的）经理；管理人员
project manager 项目经理

manually /ˈmænjuəlɪ/ *adv.* 手动地；用手
manually actuated 人工启动的

mark /mɑːrk/ *n.* 污点；标志；记号；分数 *v.* （给学生或功课）打分；做标记；标示；纪念；标志（重要事件或时刻）；（特征，特点）标志；留心；盯住（对手）；分隔
trade mark 商标

marking /ˈmɑːrkɪŋ/ *n.* 脏污，留下痕迹；标识 *v.* 做记号
marking pen 记号笔；marking tool 画线工具

marking sheet 答题纸

match /mætʃ/ *v.* 匹配；使相配；与……一致；使较量；适应 *n.* 比赛；火柴；相配的人（或物）；配偶；匹配

perfect match 完美组合；The Little Match Girl 卖火柴的小女孩

maximum / ˈmæksɪməm / *adj.* 最高的；最多的；最大极限的 *n.* [数] 极大，最大限度；最大量

maximum number 最大数；maximum limit 最大限度；maximum temperature 最高温度

measure / ˈmeʒər / *v.* 测量 *n.* 措施，办法

technical measure 技术措施

measurement / ˈmeʒərmənt / *n.* 测量（评分）；三围；计量；测量单位

unit of measurement 计量单位；measurement data 测量数据；measurement error 测量误差

measurement equipment 测量仪器

medal / ˈmed(ə)l / *n.* 勋章，奖章；纪念章

gold medal 金牌；silver medal 银牌；bronze medal 铜牌；medal of honor 荣誉勋章

merge / mɜːrdʒ / *v.* （使）合并；融入；兼并（产权，产业）

merge together 混合起来

micrometer / maɪˈkrɑːmɪtər / *n.* 千分尺

depth micrometer 深度千分尺

mixed ink 调配好的（专色）油墨

monitor / ˈmɑːnɪtər / *n.* 监控器；显示器；班长 *v.* 监视；监听（外国广播或电话）

system monitor 系统监视器

monitor quality 监控质量

mount / maʊnt / *v.* 安装（印版）；组织；增多，上升；增强

mount up 增长；上升

N

national / ˈnæʃ(ə)nəl / *adj.* 国家的；国民的；民族的；国立的 *n.* 国民

national economy 国民经济；national standard 国家标准；national security 国家安全；National Day 国庆节

navigation / ˌnævɪˈgeɪʃ(ə)n / *n.* 航行；航海

satellite navigation 卫星导航；navigation bar 导航栏；navigation equipment 导航设备

non-image area 非图像区域

normal / ˈnɔːrm(ə)l / *adj.* 正常的；正规的；标准的 *n.* 正常；标准；常态

normal university 师范大学；normal operation 常规操作；normal temperature 正常体温

normally / ˈnɔːrməli / *adv.* 正常地；通常地，一般地

normalization *n.* 标准化

notify / ˈnoʊtɪfaɪ / *v.* 示意；通报，告知；申报

notify party 通知方

number of attempts 尝试（解决问题的）次数

number of copies 印张数量

O

occur / əˈkɜːr / v.（尤指意外地）发生；出现；存在
occur as 以……的形式出现；occur for 发生在……时候

official / əˈfɪʃ(ə)l / adj. 官方的；正式的；公务的 n. 官员；公务员
official residence 官邸；official language 官方语言；official website 官方网站

offset / ˈɔːfset / n. 胶印 adj. 胶印的，平版印刷的
offset printing 胶印

operator / ˈɑːpəreɪtər / n.（设备、机器的）操作员；经营者
system operator 系统操作员

order of cutting 裁切顺序

organize / ˈɔːrɡənaɪz / v. 组织；安排；照料
organize my thoughts 整理思绪

orient / ˈɔːrɪent / v. 朝向；确定方位；引导；使熟悉 n.＜旧＞东方，东亚诸国（the Orient）；优质珍珠 adj.＜文＞东方的；（尤指宝石）光彩夺目的；（太阳等）冉冉上升的
Orient Express 东方快车；Pearl of the Orient 东方之珠

oversee / ˌoʊvərˈsiː / v. 监督；审查；俯瞰；偷看到，无意中看到
oversee the implementation of... 监督执行……

overwhelming / ˌoʊvərˈwelmɪŋ / adj. 压倒性的；势不可挡的
overwhelming superiority 绝对优势

P

pace / peɪs / n. 一步；步速；步伐；（移动的）速度 v.（因担忧、紧张或不耐烦而）踱步；踱步于；缓慢而行
a change of pace 节奏变换；keep pace with 保持同步；to set the pace 控制进攻节奏；pace maker 领跑

packing / ˈpækɪŋ / n.（橡皮布）包衬；包装；填充物 v. 包装
packing list 包装单

package / ˈpækɪdʒ / n.（橡皮布）包衬；包裹；（建议或提供的想法或服务的）一套 v. 把……打包；包装（产品或想法）
package size 包装尺寸

page / peɪdʒ / n.（书、报纸、文件等的）页，面
page size 页面尺寸
page list 页面列表

panic / ˈpænɪk / n. 恐慌，惊慌；大恐慌 adj. 恐慌的；没有理由的 v.（使）恐慌
panic attack 恐慌发作

Pantone / pænton / *n.* 潘通色卡
PANTONE FORMULA GUIDE 彩通配方指南

Pantone book 潘通色卡

Pantone ink 潘通油墨

paper cutter 切纸机

paper size 纸张规格

participant / pɑːˈtɪsɪpənt / *n.* 参与者，参加者 *adj.* 参与的
participant observation 参与观察

participate / pɑːˈtɪsɪpeɪt / *v.* 参与，参加；分享
public participation 公众参与；participate in sth. 参加某事

perimeter / pəˈrɪmɪtər / *n.* 周长；周界
security perimeter 安全边界

permanent / ˈpɜːrmənənt / *adj.* 永久的；（尤问题或困难）一直存在的
permanent marker 永久记号笔；permanent resident 永久居民

phobia / ˈfoʊbɪə / *n.* 恐怖，憎恶；恐惧（症）
silence phobia 怕安静

physical / ˈfɪzɪk(ə)l / *adj.* 物理的；身体的 *n.* 体格检查
physical object 物体

pin / pɪn / *n.* 徽章；（尤指做衣服时固定布料用的）大头针，别针 *v.* （用钉子）钉住；压住；将……用针别住
rolling pin 擀面杖；pin something on someone [口语] 把某事的责任加在某人身上；on pins and needles 如坐针毡；pin down 使受约束；阻止

place in 放置

plate / pleɪt / *n.* 印版；盘子，碟子；一盘（食物）
fruit plate 水果盘

Portuguese / ˌpɔːrtʃʊˈɡiːz / *adj.* 葡萄牙的；葡萄牙人的 *n.* 葡萄牙语；葡萄牙人

position / pəˈzɪʃn / *n.* 位置；地点 *v.* 安置；为（产品，服务，业务）打开销路
social position 职位；社会地位

preflight / priːˈflaɪt / *adj.* 预飞；起飞前的
preflight checks 飞行前检查

preparation / ˌprepəˈreɪʃ(ə)n / *n.* 预备；准备（指动作或过程）
surface preparation 表面处理；preparation for 建议

pre-set 预设

press / pres / *n.* 印刷机；新闻工作者；新闻报道 *v.* 压，挤，推
printing press 印刷机

press speed 印刷速度

pressure / ˈpreʃər / *n.* 压力；压迫，压强 *v.* 迫使；密封；使……增压
under pressure 面临压力；high pressure 高压；blood pressure 血压

Prinect 印通

printability / ˌprɪntəˈbɪlɪtɪ / *n.* 印刷适性，可

印染的

printability test 印刷适性测试

print house 印刷厂

process steps （工作）流程步骤

production / prəˈdʌkʃn / *n.* （以销售为目的的）生产，制造

production line 流水线

productivity /ˌprɒdʌkˈtɪvətɪ / *n.* 生产率

labor productivity 劳动生产率

program / ˈprəʊɡræm / *n.* （计算机）程序 *v.* 设置

automatic program 自动程序

pronounce / prəˈnaʊns / *v.* 发音；宣判；断言；发表意见，表态

pronounce on 对……发表意见

proof / pruːf / *n.* 校样，样张

content proof 样张

punch / pʌntʃ / *v.* （印版）打孔 *n.* 一拳，一击

punch press 打孔机；punch card 打孔卡

pursue / pərˈsuː / *v.* 继续探讨；从事；追赶；纠缠

pursue one's studies 治学；the career I pursue 我理想的事业；pursue pleasure 寻欢作乐

Q

qualification /ˌkwɑːlɪfɪˈkeɪʃn / *n.* （通过考试或学习课程取得的）资格；条件；限制；赋予资格

qualification certificate 资格证书；professional qualification 专业资格

quality / ˈkwɑːlətɪ / *n.* 质量 *adj.* 优质的，高质量的

quality control 质量控制；high quality 高品质

quantity / ˈkwɑːntətɪ / *n.* 量，数目

quantity production 批量生产

R

railway / ˈreɪlweɪ / *n.* <英>铁路；<英>（其他交通工具使用的）轨道；铁道部门 *v.* 乘火车旅行

railway station 火车站

raise / reɪz / *v.* 提高（数量、水平等）；筹集；养育；升起；饲养，种植 *n.* 高地；上升；<美>加薪

raise oneself 长高；raise money 集资；募捐；get a raise 得到加薪；raise capital 筹集资本

range / reɪndʒ / *n.* （变动或浮动的）范围；幅度；排；山脉 *v.* （在一定范围内）变动；漫游；射程达到；使并列

a range of 一系列；full range 全距；in the range of 在……范围之内

raster / ˈræstər / *n.* （电子）光栅；试映图

Raster Image Processor（RIP）光栅图像处理器

receptive / rɪˈseptɪv / *adj.* 可以接受的；（对观点、建议等）愿意倾听的
receptive unit 感觉单位

recognition / ˌrekəɡˈnɪʃ(ə)n / *n.* 识别；承认
recognition system 识别系统；speech recognition 语音辨识

reference / ˈrefrəns / *adj.* 参考的；文献索引的，参照的 *n.* 提及；参考；引文；（帮助或意见的）征求；推荐信；介绍人 *v.* 提及；引用
reference sample 参考样张

registration / ˌredʒɪˈstreɪʃ(ə)n / *n.* 套准；定位；登记，注册；挂号
registration form 注册表

remaining sheets 剩余纸张

remove / rɪˈmuːv / *v.* 搬出；移开；废除；把……免职；脱下
remove grain 移除颗粒；remove all 全部删除

represent / ˌreprɪˈzent / *v.* 代表；表现；描绘；回忆；再赠送；提出异议
represent something to oneself 想象出某事物；represent for 代表；象征

reprint / ˈriːprɪnt / *v.* 重印；再版 *n.* 重印；翻版
reprint editon 重印版

resolution / ˌrezəˈluːʃn / *n.*（电视、照相机、显微镜等的）分辨率，清晰度；正式决定；（冲突、问题等的）解决办法；决心
image resolution 图像分辨率；high resolution 高分辨率；dispute resolution（法律）调解纠纷；conflict resolution 冲突解决；low resolution 低分辨率

reverse / rɪˈvɜːrs / *adj.* 相反的；反向的 *v.* 逆转（决定、政策、趋势等）；撤销（法庭判决）；颠倒；交换（位置、功能）*n.* 逆向，逆转
reverse psychology 逆反心理

roller / ˈroʊlər / *n.* 辊；滚动（或碾压）东西的人
roller coaster 过山车
roller clean-up blade 辊清洁刮刀
roller stripe gauge 辊压痕量规

roster / ˈrɑːstər / *n.* 花名册；执勤人员表
duty roster 轮值表；shift roster 轮班

routine / ruːˈtiːn / *n.* 常规，惯例 *adj.* 常规的，例行的 *v.* 按惯例安排
routine business 日常事务；utility routine 实用程序；daily routine 日常生活；required routine 规定动作

S

scheme / skiːm / *n.*<英>（政府或其他组织的）计划；组合；体制；诡计 *v.* 搞阴谋；

拟订计划；策划

pension scheme 退休金计划；sampling scheme 抽样方案；color scheme 配色方案

saddle /ˈsæd(ə)l/ *n.* 鞍，马鞍 *v.* 使负重担；跨上马鞍

saddle stitched 骑马订；bicycle saddle 自行车座

safely install 安全安装

safety /ˈseɪftɪ/ *n.* 安全；安全性；安全场所 *adj.*（特征、措施）保障安全的

safety goggle 安全眼罩；safety valves 安全阀

sample /ˈsæmpl/ *n.* 样张；样本，样品 *v.* 品尝；体验（活动）；对……作抽样调查

sample size 样本尺寸

satin /ˈsætn/ *n.* 缎子；缎子衣服 *adj.* 光滑的

satin grain long paper 光面长丝缕纸张

scale /skeɪl/ *n.* 磅秤；等级；刻度；规模；比例；鳞；音阶 *v.* 改变（文字、图片的）尺寸大小；刮去（鱼鳞）；攀登；剔除（牙垢）；称得重量为 *adj.*（模型或复制品）按比例缩小的

ink scale 油墨秤；economies of scale 规模经济；large scale 大规模；on the scale of 按……的比例；scale up 按比例放大

scanning /ˈskænɪŋ/ *n.* 扫描

color scanning 色彩扫描

schedule /ˈskedʒuːl/ *n.* 计划（表）；＜美＞（公共汽车、火车等的）时间表；一览表 *v.* 安排，预定；将……列入计划表或清单

on schedule 按时；ahead of schedule 提前；according to schedule 按照预定计划；production schedule 生产计划

scratch /skrætʃ/ *n.*（某人皮肤上的）划痕；抓，挠 *v.*（用指甲）挠，轻抓；划出（痕迹）；（用爪子）刨；勾掉（写下的字）*adj.*（球队或一群人等）仓促拼凑的，匆匆组成的

start from scratch 从零开始；cat-scratch disease 猫抓病

scumming /ˈskʌmɪŋ/ *n.* 起脏；撒渣；吐渣 *v.* 除去（浮渣）

de-scumming 去渣

security /sɪˈkjʊrətɪ/ *n.* 安全，安全性；保证；证券；抵押品

network security 网络安全；Security Bureau 安全局；homeland security 国土安全；Security Council 安理会；security guard 保安员

sequence /ˈsiːkwəns/ *n.* 顺序；连续事件（或动作、事物）*v.* 按顺序排列；测定（整套基因或分子成分的）序列；用音序器播放（或录制）音乐

sequence folder 序列文件夹

sequence order 顺序

server /ˈsɜːrvər/ *n.*（电脑）服务器

virtual server 虚拟服务器

settle / ˈset(ə)l / v. 解决（分歧，纠纷等）；定居；沉淀；（地面或建筑）下陷 n. 有背长椅

settle down 定居；settle down to 专心致力于；settle up 付清

set up 设置

set-up waste 校版纸

sheet / ʃiːt / n. 纸张

OK sheet OK 样；sheet size 纸张规格；make ready sheet 校版纸

sheetfed / ʃiːt ˈfed / adj. 单张印刷的

sheetfed offset 单张纸胶印（机）

shuttle / ˈʃʌt(ə)l / n. 航天飞机（space shuttle 的简称）；穿梭班机、公共汽车等 v. 穿梭，往返

space shuttle 航天飞机；shuttle bus 班车

signature / ˈsɪɡnətʃər/ n.（印刷）折标；签名；签署；鲜明特色，明显特征

signature size 签样尺寸

silver / ˈsɪlvər / n. 银；银器（尤指餐具）adj.（有关）银的；含银的；<美>口才流利的 v. 变成银色

silver lining（不幸或失望中的）一线希望

simultaneously / ˌsaɪm(ə)lˈteɪnɪəslɪ / adv. 同时地

develop simultaneously 同时发展

simulation / ˌsɪmjuˈleɪʃ(ə)n / n. 模拟，仿造物；假装，冒充

simulation game 模拟游戏；simulation test 模拟试验

smash / smæʃ / v. 撞毁（交通工具）；打碎；（使）猛撞；（轻松）打破（纪录）n. 猛烈撞击声；猛击 adv. 哗啦一声 adj. 非常轰动的，出色的

smashed blanket 橡皮布塌陷；smash shot 扣杀球

software / ˈsɔːftwer / n. 软件

software development 软件开发；software license 软件许可证

specific / spəˈsɪfɪk / adj. 特定的；具体的；有特效的；n. 细节；特效药

specific instructions 明确的指示

specific characteristic 特定特征

specification / ˌspesɪfɪˈkeɪʃ(ə)n / n. 规格；规范；说明书；详述

technical specification 技术规范；design specification 设计规格

speck / spek / n. 污点；灰尘；小颗粒 v. 使有斑点

paper speck 纸斑

spectrodensitometer / ˌspektrədensɪˈtɑːmɪtər / n. 分光光度仪

Double-beam spectrodensitometer 双光束分光光度仪

SPH 张/小时，sheets per hour 的缩写

spot / spɑːt / n. 斑点；地点；（人体的）部位；污渍；（皮肤上的）丘疹；排名位

置；几滴（液体）；困境；现金交易；（人格或名誉的）污点；聚光灯（spotlight 的简称）v. 看见；注意到；（对比赛对手）让分；使有污迹；将（台球）放在置球点上 adj. 现货交易的，立即支付的

color spots 色块；spot color 专色；on the spot 当场；color bar spot 色带（上的）色块

stack / stæk / n. 堆；大量，许多

a stack of 一堆……

stadium / ˈsteɪdɪəm / n.（品质的）体育场；露天大型运动场

indoor stadium 室内运动场

standard / ˈstændərd / n.（品质的）标准，水平，规范 adj. 普通的，标准的

standard error 标准误差

status / ˈsteɪtəs; ˈstætəs / n. 状况，情形

job status 工作状态

sticker / ˈstɪkər / n. 不干胶；（有图或文字的）粘贴标签，贴纸

bar coded sticker 条形码贴纸

stock / stɑːk / n. 纸张；（商店的）库存；储备物；储备量；股本；股票；家畜 v.（商店或工厂）储备；为……备货 adj. 老一套的；（商店里）库存的

stock market 股市；stock exchange 证券交易所

storage / ˈstɔːrɪdʒ / n.（信息的）存储；仓库；贮藏所

data storage 数据存储；storage system 存储系统；energy storage 蓄能；storage tank 储油罐

straight / streɪt / adj. 直的；连续的；正直的；整齐的 adv. 直接地；不断地 n. 直线

go straight 改过自新；straight up 直率地；straight line 直线；straight out 直言地

strictly / ˈstrɪktlɪ / adv. 严格地；完全地；确实地

strictly confidential 绝对机密

style / staɪl / n.（字）形；方式；款式 v. 设计，给……造型；称呼，命名

text style 字体样式

suspension / səˈspenʃ(ə)n / n. 悬浮；暂停；停职

in suspension 悬浮中

swatch / swɑːtʃ / n. 样本，样品

sample swatch 样张

T

target / ˈtɑːrgɪt / n. 目标，指标；（攻击的）对象；靶子 v. 把……作为目标；面向，对准（某群体）

target market 目标市场；on target 切中要害

task / tæsk / n. 任务 v. 派给……任务

task analysis 任务分析

tasks to be done 必要工作

technician / tekˈnɪʃn / n. 技术人员，技师

X-rite technician 爱色丽技术人员

template / ˈtemplət / n.（计算机文档的）模板；样板；垫木

 template parameter 模板参数；web template 网页模板

terabyte / ˈterəbaɪt / n.（计算机）万亿字节，兆兆字节（信度量度单位）

 one terabyte 1TB

text / tekst / n.（书、杂志等中区别于图片的）正文，文字；文字材料 v.（用手机）给……发短信

 text editor 文本编辑器；text style 字体样式

thickness / ˈθɪknəs / n. 厚度，粗细；浓度，密度；最厚（或最深）处 v. 刨，削（木头至合适的尺寸）

 film thickness 漆膜厚度

tie / taɪ / v. 打结；连接；与……成平局 n. 领带；领结；束缚；系梁；平局

 tie in 使结合；使配合得当；tie with 在比赛中得分与……相同；tie up 占用；tie in with 与……一致；配合

timed event 计时工作

title / ˈtaɪt(ə)l / n. 标题；（人名前表示地位、职业或者婚姻状况的）头衔；（电影、电视的）字幕 v.（给书籍、乐曲等）加标题；赋予头衔；把……称为 adj. 冠军的；标题的；头衔的

 title bar 标题栏

tool / tu:l / n.（尤指手用）工具；蠢人；（书籍装订时的）压印图案 v. 驱车兜风；（用工具）制作，在（皮革，尤指书籍的皮革封面）上压印图案

 power tool 电动工具

top of the stack 收纸堆顶部

total amount of paper 纸张总数

tough / tʌf / adj. 艰苦的；坚强的；强壮的 n. 恶棍 v. 坚持；忍受，忍耐 adv. 强硬地，顽强地

 tough guy 硬汉

traditional / trəˈdɪʃ(ə)l / adj. 传统的；（活动）惯例的

 traditional Chinese medicine 中医

transfer / trænsˈfɜ:r / v. 转印（图画，图案）；（使）转移；传染（疾病），转让（权力等）n. 转移，调动；（旅行中）转乘；<美>换乘票；纸上可转印的图画或图案

 technology transfer 技术转移；wire transfer 电汇

trend / trend / n. 趋势，倾向；走向 v. <美>趋向，伸向

 market trend 市场趋势；general trend 一般趋势；trend analysis 趋势分析

trim / trɪm / v. 裁切；修剪；削减；修饰（尤指某物的边缘）n.（尤指毛发的）修剪；额外装饰 adj. 整齐的；修长的

 trim marks 裁切标记

trim line 裁切线

trim size 裁切尺寸

trolley / ˈtrɑːli / *n.* 收纸台板 *v.* 用手推车运
    trolley case 拉杆箱

turn in 上交

turn on 打开

two color press 双色印刷机

U

unit / ˈjuːnɪt / *n.* （印刷机）单元；单位
    unit of time 单位时间

unused / ʌnˈjuːzd; ʌnˈjuːst / *adj.* 不用的；从未用过的
    unused land 荒地

V

value / ˈvæljuː / *n.* 价值；等值
    target density values 目标密度值

variable / ˈveriəb(ə)l / *adj.* 可变的；易变的；时好时坏的 *n.* 可变性，可变因素
    variable data 可变数据；control variable 控制变量

VDP 可变数据印刷，Variable Data Printing 的缩写

verification / ˌverɪfɪˈkeɪʃn / *n.* 验证；核查；证实
    quality verification 质量检验；ink verification 油墨验证

vocational / vəʊˈkeɪʃən(ə)l / *adj.* 职业的，行业的
    vocational training 职业培训；vocational education 职业教育；vocational school 职业学校

W

wipe / waɪp / *v.* （用布、手等）擦干净；解雇（某人）；刷（卡）*n.*（湿）抹布；擦
    wipe off 擦掉；还清；wipe out 擦净

work area 工作区域

workflow / ˈwɜːrkfloʊ / *n.* 工作流程
    engineer workflow 工程师工作流程

workspace / ˈwɜːrkspeɪs / *n.* 工作区域
    close workspace 关闭工作区

work station 工作站